Risk Management of
Medical Devices

Risk Management of Medical Devices

Shalinee Naidoo

ARCLER
PRESS

www.arclerpress.com

Risk Management of Medical Devices

Shalinee Naidoo

Arcler Press
4164 Lakeshore Road
Burlington, ON L7L 1A4
Canada
www.arclerpress.com
Email: orders@arclereducation.com

ISBN: 978-1-77956-469-6 (Hardcover)

ABOUT THE AUTHOR

Shalinee Naidoo is currently the Regulatory and Product Development Manager of a medical device manufacturer based in South Africa. She is directly involved in global regulatory compliance and the design and development of new medical devices from idea conception to market. She is also the founder of Life of Shal (www.lifeofshal.com) an online travel journal created to inspire others to explore the world and Scientist's Sanctuary (www.scientistssanctuary.com) – a science communication company that specializes in bridging the gap between scientific knowledge and creative communication for both the academic and corporate world.

TABLE OF CONTENTS

LIST OF FIGURES

LIST OF TABLES

PREFACE

The journey of creating and managing medical devices is a complicated picture of innovation, regulation, and risk management. These devices, which range from simple bandages to life-sustaining equipment, hold the promise of enhancing and extending lives, yet their development and deployment carry inherent risks. The stakes are extraordinarily high, as the lives and well-being of patients depend on the meticulous processes involved in mitigating these risks.

This book, *Risk Management of Medical Devices*, serves as a comprehensive guide to understanding, navigating, and implementing risk management strategies in the medical device industry. It is structured to provide readers with both foundational knowledge and practical applications, ensuring a thorough grasp of the principles and standards that underpin this vital field.

The globalization of the medical device industry has led to an unparalleled surge in innovation, but it has also introduced new dimensions of complexity and risk. Chapter 1, Introduction, establishes the groundwork by exploring the evolution of medical devices, their classification, and the multifaceted nature of risk. By unpacking concepts such as business, financial, environmental, and health risks, this chapter illuminates the interconnections between these categories and underscores the importance of holistic risk management.

Chapters 2 and 3 examine the methodologies that drive effective risk management. Techniques such as Preliminary Hazard Analysis (PHA), Failure Mode and Effects Analysis (FMEA), Fault Tree Analysis (FTA), and Hazard Analysis and Critical Control Point (HACCP) are presented with clarity and depth. These chapters offer actionable insights into choosing and implementing these tools, complete with discussions on their respective advantages and limitations.

Chapter 3 expands on the risk management process itself, detailing essential definitions, stages, and frameworks. Readers are guided step-by-step through the lifecycle of risk management, from hazard identification to the evaluation of residual risks, empowering professionals to adopt systematic and effective approaches.

Chapters 4 and 5 are dedicated to the standards and documentation that form the backbone of risk management in the medical device industry. Standards like ISO 14971, ISO 24971, and IEC guidelines are demystified, offering readers a clear understanding of their purposes, principles, and applications. In an industry governed by stringent regulatory expectations, this knowledge is indispensable.

The chapters further explore how standards evolve, how they interrelate, and how they shape regulatory compliance. The focus on ISO 14971, in particular, provides a lens into its application, highlighting the differences across its 2007, 2012, and 2019 iterations, and demonstrating how it harmonizes with other standards.

Risk management is not a one-time task but an ongoing process, as emphasized in Chapters 5 and 6. These sections address the creation and maintenance of a Risk Management File, a crucial element for both compliance and operational safety. Templates and best practices are included to guide readers through the development of plans, assessments, and reviews. Moreover, the chapters stress the importance of post-production monitoring and data analysis, ensuring continuous improvement and vigilance.

To enhance practical understanding, Chapter 7 provides templates and annexures, including risk policy and management plans, making it easier for professionals to apply theoretical knowledge in real-world scenarios. Chapter 8 expands the reader's toolkit by introducing additional standards that complement ISO 14971 and ISO 24971.

This book is written for a diverse readership that spans professionals and stakeholders in the medical device industry. It is an essential resource for medical device designers, manufacturers, quality assurance professionals, regulatory affairs specialists, and healthcare providers who are directly involved in the lifecycle of medical devices. Additionally, it serves as a valuable guide for academic researchers, engineering students, and policymakers seeking to understand the intricacies of risk management in a highly regulated and impactful industry. With its clear explanations of standards, practical methodologies, and compliance strategies, this book is written to bridge knowledge gaps and empower readers to navigate the complexities of medical device safety and innovation with confidence.

CHAPTER 1

INTRODUCTION

Contents

1.1. OVERVIEW OF MEDICAL DEVICE GLOBALIZATION AND RISK MANAGEMENT

Medical devices play a critical role in modern healthcare delivery. The globalization of medical devices, coupled with increasing demand and usage, has amplified the need for the design and development of safe devices for human use. In the year 2000, it was estimated that approximately 1.5 million different medical devices were available on the global market. As innovation and technological advancements rapidly progress, medical devices have become one of the fastest-growing industries worldwide. A major challenge for medical device manufacturers is meeting the stringent safety and compliance requirements necessary for approval to ensure devices can be safely used on patients. One of the fundamental safety requirements is the management of unavoidable risks associated with the design and use of the device.

Risk is often defined as the combination of the probability of occurrence and the severity of harm that can arise from the use of a medical device. The process of implementing risk controls for medical devices is complex due to the wide range of devices available and the various stages in their lifecycle. The risks associated with each device type can vary significantly, making the potential for misuse substantial. Therefore, all medical devices must undergo comprehensive risk control measures to ensure safety.

Risk management has become a crucial tool that, when executed correctly, can demonstrate compliance with regulatory requirements and facilitate market access in various regions. Many stakeholders, including clinicians, patients, manufacturers, regulators, and litigators, are becoming increasingly aware of human-related safety issues. Risk management involves systematically applying various approaches, procedures, and practices to assess, evaluate, and control risks. While eliminating risk entirely is not possible, it is feasible to limit risks if they are identified and properly managed. Additionally, device usability, safety, and compliance must be carefully considered.

Standards are essential for establishing appropriate methodologies for risk management. One such standard is ISO 14971, which will be explored in detail in this book. However, it is important to note that ISO 14971 is not the only risk management standard, and other standards touching on risk management will also be discussed.

1.1.1. What Is a Medical Device?

Before delving into risk and risk management strategies, it is important to define what constitutes a medical device. A medical device is any product or equipment designed for use in the healthcare sector with the purpose of:

- Diagnosing, monitoring, or treating diseases; or
- Serving as a supportive aid for individuals with illnesses or disabilities.

The **World Health Organization (WHO)** defines a medical device as:

> "*Any instrument, apparatus, implement, machine, appliance, implant, reagent for in vitro use, software, material, or other related article intended by the manufacturer to be used alone or in combination for human beings for one or more specific medical purposes.*"

General uses for medical devices, as outlined by the WHO, include:

- Diagnosis, prevention, monitoring, treatment, or alleviation of diseases;
- Diagnosis, monitoring, treatment, alleviation, or compensation for injuries;
- Investigation, replacement, modification, or support of the anatomy or physiological processes;
- Supporting or sustaining life;
- Control of conception;
- Disinfection of medical devices;
- Providing information through in vitro examination of specimens derived from the human body.

Notably, medical devices do not achieve their primary intended function through pharmacological, immunological, or metabolic means. However, these means may assist the device in fulfilling its purpose. This distinction separates medical devices from medicines or drugs, whose primary action is often achieved through chemical interaction or metabolism in the body.

The Global Harmonization Task Force (GHTF) has proposed a harmonized definition of medical devices and in vitro medical devices to standardize understanding globally. According to the GHTF document SG1/N029R11, a medical device is defined as:

> "Any instrument, apparatus, implement, machine, appliance, implant, in vitro reagent or calibrator, software, material, or other similar or related article, intended by the manufacturer to be used, alone or in combination, for human beings for one or more specific purposes of:
>
> - Diagnosis, prevention, monitoring, treatment, or alleviation of disease;

- Diagnosis, monitoring, treatment, alleviation of or compensation for an injury;
- Investigation, replacement, modification, or support of the anatomy or a physiological process;
- Supporting or sustaining life;
- Control of conception;
- Disinfection of medical devices;
- Providing information for medical purposes through in vitro examination of specimens derived from the human body;
- Which does not achieve its primary intended action in or on the human body by pharmacological, immunological, or metabolic means, but which may be assisted in its function by such means."

The GHTF also defines an **In-Vitro Medical Device** (GHTF/SG1/N071:2012) as:

> "*A medical device, whether used alone or in combination, intended by the manufacturer for in-vitro examination of specimens derived from the human body, primarily to provide diagnostic, monitoring, or compatibility information.*"

1.2. CLASSIFICATION OF MEDICAL DEVICES

Medical devices vary widely, from simple low-risk products like tongue depressors and latex gloves to complex devices such as pacemakers or software used for medical purposes. According to ISO 13485, medical devices are classified based on the level of risk associated with their use. Classification systems differ by region, but the goal is to ensure devices are subjected to appropriate regulatory controls.

In Europe, medical devices are classified into four classes:

- **Class I:** Lowest risk;
- **Class IIa;**
- **Class IIb; and**
- **Class III:** Highest risk.

In South Africa, the classification system uses:

- **Class A:** Lowest risk;
- **Class B**;
- **Class C**; and

- **Class D**: Highest risk.

The following diagram provides a visual representation of this classification system:

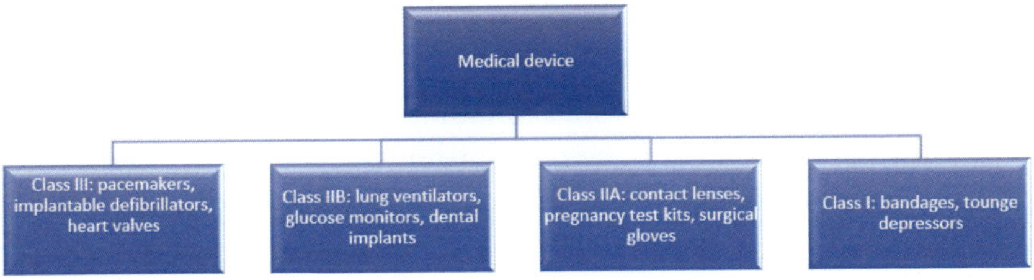

1.3. WHAT IS RISK?

Risk is defined as the chance of something happening that could have a negative effect. In simplified terms, a risk is the possibility of something bad happening. It often involves uncertainty about the effects of an activity concerning something that is valued, such as health, well-being, property, or the environment. While many definitions have been proposed, the internationally standardized definition of risk is "the effect of uncertainty on objectives."

It is also important to note that the descriptions of risk, the definitions of risk, and the overall method of assessment and management of risk differ depending on the area of practice, which could be anything from business to the environment, finance, healthcare, insurance, security, and so on.

There are many definitions of risk that have been put forth to date. The Oxford English Dictionary cites the earliest use of the word as from 1621, along with many other definitions. Briefly, the word 'risk' is defined as:

> *(Exposure to) the possibility of loss, injury, or other adverse or unwelcome circumstance; a chance or situation involving such a possibility.*

The **International Organization for Standardization (ISO) Guide 73** (2009) on risk management, which has now been withdrawn, also defined risk. The definition below was developed by an international committee representing over 30 countries and is based on the input of various experts:

> **Effect of uncertainty on objectives**

The following notes were further provided:

- **Note 1:** An effect is a deviation from the expected – positive or negative.
- **Note 2:** Objectives can have different aspects (such as financial,

health and safety, and environmental goals) and can apply at different levels (such as strategic, organization-wide, project, product, and process).

- **Note 3:** Risk is often characterized by reference to potential events and consequences or a combination of these.
- **Note 4:** Risk is often expressed in terms of a combination of the consequences of an event (including changes in circumstances) and the associated likelihood of occurrence.
- **Note 5:** Uncertainty is the state, even partial, of deficiency of information related to, understanding or knowledge of, an event, its consequence, or likelihood.

Risk is present in various areas, but it is important to note that the measurement of risk and even the definitions of risk differ from one practice area to another. Below are a few different types of risk in more detail.

1.3.1. Business Risk

Business risks arise from uncertainties related to the profitability of a business. This could be due to unwanted events, changing customer preferences, increased or changing competition, changes in government policy, and various other factors.

1.3.2. Economic Risk

Economic risk arises from the uncertainty about economic outcomes. This could be due to exchange rates, government regulations, or political stability/instability, all of which could affect a company's prospects.

1.3.3. Environmental Risk

Environmental risk refers to potential hazards to the environment and human health, such as air and water pollution. These risks arise from environmental hazards or issues and could have harmful effects on human health, surrounding environments, or ecological systems. Environmental risks can impact health directly in various ways, such as causing asthma or allergies. Environmental risk factors include pollution, radiation, noise, land use patterns, and climate change, to name a few. Air pollution is often the main environmental risk factor.

1.3.4. Financial Risks

Financial risks arise from uncertainty about financial returns and involve risks related to markets, liquidity, and various other operational and credit risks. The risk in this discipline is that the actual return on an investment will differ from its expected/projected return. This could be negative or positive, with returns below expectations, including the possibility of losing some or all of the original investment, or returns exceeding expectations.

1.3.5. Health Risks

A health risk is the chance or likelihood that something will harm or affect personal human health. Personal human health includes factors such as sex, family health history, lifestyle, and so on. This type of risk arises from disease or any form of biological hazard and involves characterizing the nature and likelihood of a harmful effect on individuals or populations from certain human activities.

1.3.6. Health, Safety, and Environmental Risk (HSE Risk)

Health, safety, and environmental (HSE) risks are separate practice areas but are often linked due to organizational management structures because a single risk event may impact all three areas. However, these could occur over different timescales, ranging from immediate or short-term effects to much longer-term environmental impacts.

1.3.7. Information Technology Risks

Risks in information technology arise from the potential for a threat to exploit a vulnerability, breach security, and cause harm. General IT risks include:

- Hardware and software failures;
- Malware, involving malicious software designed to disrupt computer operations;
- Viruses, which may spread from one computer to another, disrupting operations;
- Spam, scams, and phishing, which involve unsolicited contact to fool people into giving personal details or buying fake goods;
- Human error, which includes accidentally opening emails containing viruses, incorrect data processing, or careless data disposal;
- Natural disasters such as floods, storms, and bushfires, which may interrupt service within the business or with external suppliers.

Managing these IT-related risks can include securing computers, servers, and wireless networks; using anti-virus and anti-spyware protection and firewalls; updating software to the latest versions; using data backups that include off-site or remote storage; securing user passwords; implementing two-factor authentication; training staff in IT policies and procedures; and understanding legal obligations for online business.

Medical device manufacturers often follow a highly structured and rigorous risk management process. This process is typically regulated by the territories in which their devices are sold, in addition to the International Standards Organization. A variety of risk management solutions are used by manufacturers to track risk throughout the entire product lifecycle. These include Failure Mode and Effects Analysis (FMEA), 5-Why, and Hazard Analysis, to name a few.

In the field of medical devices, risk is considered in the context of the safety, effectiveness, and performance of the device in question and in relation to the end user while taking various standards and regulatory requirements into consideration. Many of these standards and regulations place greater emphasis on a risk-based quality management system (QMS). This covers not only risks associated with the product lifecycle but also other areas such as supply chain, clinical studies, usability, and post-market surveillance.

1.4. WHAT IS RISK MANAGEMENT?

Risk management is a defined process for identifying potential hazards and mitigating them. The risk management process generally starts with the design process to minimize potential hazards from the onset of development and involves the identification, understanding, control, and prevention of failures that could result in hazards when people use the device in question. The process continues through the product lifecycle and generally includes provisions for reviewing post-production data. This ensures that the device's benefits continue to outweigh the risks and that any newly identified hazards are mitigated as much as possible.

Ideally, manufacturers are expected to identify possible hazards associated with the device in both normal and fault conditions. If these risks are deemed unacceptable, they should be reduced to acceptable levels by appropriate means.

The need to perform risk management is vast and should be done for the following reasons:

- It is required by law and governed by various standards and regulations.
- It allows identification of device design problems before distribution occurs.
- It eliminates the costs associated with recalls if done correctly.
- It offers a measure of protection from product liability.

- It is required for regulatory submissions.
- It ensures the safety of the device in question.
- It ensures that unsafe devices do not reach the market without proper risk mitigation.
- It demonstrates that manufacturing has been carried out in a controlled manner and has resulted in the production of a safe device.

The key stages in a risk management process include:

- Identification of hazards;
- Evaluation of consequences and their likelihood to determine a measure of risk;
- Comparison to the company's risk acceptability criteria;
- Mitigation if needed.

This is illustrated in Figure 1.1.

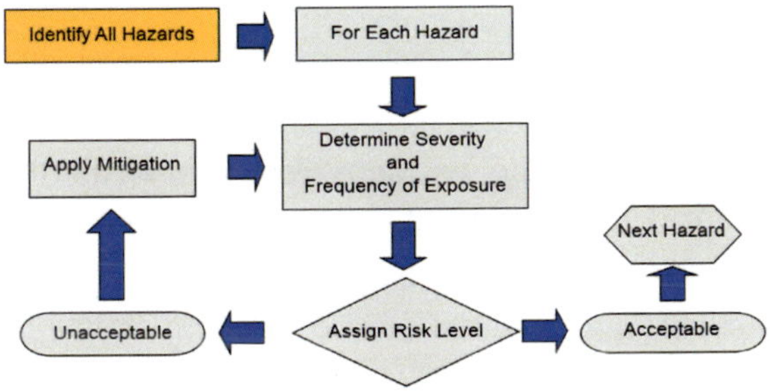

Figure 1.1. *Risk management flow chart.*

It is also important to note that in some cases, risks cannot be completely eliminated, and therefore any remaining risk must be managed and mitigated as far as possible. Often, companies will devise a risk management program to aid in managing risks. This often includes:

- The development of protocols or standard operating procedures for what needs to be done and how to do it.
- Definition of responsibilities within the organization.
- Definition of what needs to be authorized and who is responsible for authorizing and handling the process.
- Definition of the skills and knowledge necessary to implement the system.
- Definition of how training will be provided for those who do not possess such skills.

- Development and maintenance of records to demonstrate conformance to policies and procedures.
- Incorporation of measures to cross-check and verify that procedures are followed.
- Verification of the systems in place and checks to ensure they are functioning properly.

1.5. WHY SHOULD WE PERFORM RISK MANAGEMENT?

There are various reasons why risk management should be performed in the medical industry. Below are some of the main reasons:

- Performing risk management activities allows for the identification of device design problems prior to distribution, which could help mitigate the cost associated with field safety corrective actions and recalls.
- It could offer a measure of protection from product liability and related damages.
- It ensures compliance, as many regulatory submission documents now require the inclusion of a risk analysis, risk management report, risk plan, and so forth.
- It ensures the safety of the device by adequately addressing risks.
- It ensures that unsafe devices do not reach the market and are promptly identified and efficiently corrected.
- It ensures that the manufacturing process produces a safe device.

1.6. THE HISTORY OF RISK MANAGEMENT

Literature indicates that risk management began to be studied after World War II. In fact, several sources date the origin of modern risk management back to 1955–1964 (Crockford, 1982; Harrington and Neihaus, 2003; Williams and Heins, 1995). The first two academic books were published by Mehr and Hedges (1963) and Williams and Heins (1964), and the content covered pure risk management.

Risk management has long been associated with the use of market insurance to protect individuals and companies from various losses. However, in the mid-1950s, new forms of pure risk management emerged as alternatives to market insurance. During the 1960s, contingency planning activities were developed, and various risk prevention or self-protection activities against some losses were put in place. It was also during this time that coverage for work-related illnesses and accidents started appearing.

In the 1980s, companies began to consider financial management or risk portfolios, while financial institutions, including banks and insurance companies, intensified their market and credit risk management activities. Operational risk and liquidity risk management emerged in the 1990s. International regulation of risk also began in the 1990s. It was during this time that risk management models and capital calculation formulas were developed by financial institutions to protect themselves from unanticipated risks and reduce regulatory capital. Simultaneously, governance of risk management became essential, integrated risk management was introduced, and the first risk manager positions were created.

The Sarbanes-Oxley Act was introduced in the United States in 2002, establishing governance rules for companies. Stock exchanges, including the New York Stock Exchange (NYSE) in 2002, also added risk management governance rules for listed companies. The management of risk has continued to evolve ever since.

CHAPTER 2

RISK MANAGEMENT IN THE MEDICAL DEVICE INDUSTRY

Contents

2.1. RISK MANAGEMENT IN THE MEDICAL DEVICE INDUSTRY

The medical device industry is built on trust between those who design, develop, and manufacture devices, and those who depend on those devices to work safely and effectively. Imagine this from the perspective of a patient going in for any medical procedure. The patient probably thinks very little about the risks of the medical devices about to be used. Generally, the patient trusts the expertise of the clinicians. The patient seldom wonders if the products used by the clinicians are safe and have been thoroughly and rigorously tested. The patient, often unknowingly, accepts the risks of the medical device you and I design, develop, and manufacture. And this is exactly why risk management is so important to the medical device industry. You have to know that the medical devices you are involved with bringing to patients and end-users are safe.

2.2. RISK ANALYSIS TECHNIQUES

Risk analysis has become one of the most fundamental requirements where regulation of medical devices is concerned. The hazardous effects of a device must be taken into consideration before a product is approved and released onto the market. For this reason, risk analyzes should be started as early as in the design phase of a medical device. This allows for early detection of adverse events that may result in serious harm to the user or patient. Some commonly used techniques for risk analysis include:

- Preliminary hazard analysis (PHA);
- Fault tree analysis (FTA);
- Failure mode and effect analysis (FMEA);
- Hazard and operability study (HAZOP);
- Hazard analysis and critical control point (HACCP).

2.3. PRELIMINARY HAZARD ANALYSIS (PHA)

Preliminary Hazard Analysis is a method of identifying potential hazards at the very early stage of design. This method may be used where a more extensive approach is not suitable and is ideal in the analysis of systems where preliminary design is in progress. In this way, product managers or designers can then take the necessary steps to alleviate the system's potential risk. PHA is a semi-quantitative analysis that is ideally performed to:

1. Identify all potential hazards and hazardous events that may lead to an accident.

2. Rank the identified hazardous events according to their severity.
3. Identify required hazard controls and follow-up actions.

PHA may also be known as Rapid Risk Ranking or Hazard Identification.

PHA can be used in the following scenarios:

1. As an initial risk study in the early stages of a project. Take, for example, the setup of a new engineering plant. Accidents may be caused by the release of energy. The PHA identifies where energy may be released and which hazardous events may occur. In addition, it gives a rough estimate of the severity of each hazardous event. The PHA results are used to then: i. Compare main concepts, ii. Focus on important risk issues, and as iii. Input to more detailed risk analyzes.
2. As an initial step of a detailed risk analysis of a concept system or an existing system. In these instances, the purpose of the PHA is to identify those hazardous events that should be subject to a further, and more detailed, risk analysis.
3. As a complete risk analysis of a simple system. The possibility of whether or not a PHA will be a sufficient analysis will depend on the complexity of the system and the objectives of the analysis.

2.4. WHAT SHOULD A PHA CONSIST OF?

A PHA should take the following into consideration:

- Hazardous components;
- Safety-related interfaces between various system elements, including software;
- Environmental constraints, including operating environments;
- Operating, test, maintenance, built-in-tests, diagnostics, and emergency procedures;
- Facilities, real property installed equipment, support equipment, and training;
- Safety-related equipment, safeguards, and possible alternate approaches;
- Malfunctions to the system, subsystems, or software.

2.5. FOUR MAIN STEPS IN ANY PHA

There are four main steps in any PHA. These include:

1. PHA prerequisites;

2. Hazard identification;
3. Consequence and frequency estimation;
4. Risk ranking and follow-up actions.

2.6. DETERMINING THE PHA PREREQUISITES

This step involves establishing the PHA team, defining and describing the system to be analyzed, and collecting risk information from previous and similar systems. A typical PHA team may consist of the following members:

- A team leader or facilitator who has the competence and experience in the methodology being used.
- A secretary who will report the results.
- Approximately two to six team members who can provide the necessary knowledge and experience on the methodology or system being analyzed.

The number of team members participating would ideally depend on the complexity of the system and the objectives of the analysis. In some instances, some team members may only participate in certain parts of the analysis.

When defining and describing the system to be analyzed, the following should be taken into consideration:

- System boundaries (i.e., which parts should be included and which parts should not).
- System descriptions – this should include layout drawings, process flow diagrams, block diagrams, and so on.
- Use and storage of energy and hazardous materials in the system.
- Operational and environmental conditions to be considered.
- Systems for detection and control of hazards and hazardous events, emergency systems, and mitigation actions.

In addition, to ensure adequate functioning of the system, the following should be considered:

- Inputs into the system/what the system is dependent on.
- What activities are performed by the system/functions of the system.
- What services does the system provide/what are the outputs of the system.

The system should be split into manageable parts, allowing ease of identifying all hazards and events. The system can ideally be split into:

- System parts;

- Activities;
- Exposure to risk.

Results of a PHA can be documented in a template that tracks the hazard, accidental event (such as what, where, when), probable causes, contingencies/preventive actions, probability, severity, and any additional comments. A template is presented below.

PHA Template **System:** **Operating mode:** **Analyst:** **Date:**							
Ref.	**Hazard**	**Accidental Event (What, Where, When)**	**Probable Causes**	**Contingencies/Preventive Actions**	**Probability**	**Severity**	**Comments**
1.							
2.							
3.							

When conducting a PHA, all hazards and possible hazardous events must be identified. All parts of the system, operational modes, maintenance operations, safety systems, etc., should be considered and recorded. To get a complete survey of all possible hazards, it may be beneficial to use a hazard checklist. Several checklists have been published and are available in the literature. An example of a checklist is presented below:

Hazard Category	**Hazard**
Mechanical hazard	Properties of machine parts or workpieces, such as: • Shape • Relative location • Mass and stability (potential/kinetic energy) • Inadequacy of mechanical strength • Accumulation of energy inside the equipment, for example, elastic elements (springs), liquids and gases under pressure, The effects of vacuum • Crushing hazards

	• Shearing hazards • Cutting or severing hazards • Entangling hazards • Drawing-in or trapping hazards • Impact hazards • Stabbing or puncture hazards • Friction or abrasion hazards • High pressure fluid injection or ejection hazards
Electrical hazards	• Contact of persons with live parts (direct contact) • Contact of persons with parts which have become live under faulty conditions (indirect contact) • Approach to live parts under high voltage • Electrostatic phenomena • Thermal radiation or other phenomena, such as the projection of molten particles and chemical effects from short circuits, overloads, etc.
Thermal hazards	• Burns, scalds and other injuries by a possible contact of persons with objects or materials with an extreme high or low temperature, by flames or explosions and also by radiation of heat sources • Damage to health by hot or cold working environment
Thermodynamic hazards	• Over pressure • Under pressure • Over-temperature • Under-temperature
Hazards generated by noise	This may result in: • Hearing loss (deafness), other physiological disorder (e.g., loss of balance, loss of awareness) • Interference with speech communication, acoustic signals, etc.
Hazards generated by vibration	• Use of hand-held machines resulting in a variety of neurological and vascular disorders • Whole body vibration, particularly when combined with poor postures

Hazards generated by radiation	• Low frequency radiation • Radio frequency radiation • micro waves • Infrared, visible and ultraviolet light • X and gamma rays • Alpha, beta rays, electron or ion beams, neutrons • Lasers
Hazards generated by materials/substances	• Hazards from contact with or inhalation of harmful fluids, gases, mists, fumes, and dusts • Fire or explosion hazard • Biological or microbiological (viral or bacterial) hazards • Flammables (ignition, fire, explosion/detonation) • Chemicals (toxicity, corrosion, o-specification) • Pollutants (emissions, effluents, ventilation)

Common sources of hazards include:

- Sources and propagation paths of stored energy in electrical, chemical, or mechanical form;
- Mechanical moving parts;
- Material or system incompatibilities;
- Nuclear radiation;
- Electromagnetic radiation (including infra-red, ultraviolet, laser, radar, and radio frequencies);
- Collisions and subsequent problems of survival and escape;
- Fire and explosion;
- Toxic and corrosive liquids and gases escaping from containers or being generated as a result of other incidents;
- Deterioration in long-term storage;
- Noise, including sub-sonic and supersonic vibrations;
- Biological hazards, including bacterial growth in such places as fuel tanks;
- Human error in operating, handling, or moving near equipment of the system;
- Software errors that can cause accidents.

While there are various methods to identify hazards, some common methods are listed below:

- Examine similar existing systems;
- Review previous hazard analyzes for similar systems;
- Review hazard checklists and standards;
- Consider energy flow through the system;
- Consider inherently hazardous materials;
- Consider interactions between system components;
- Review operation specifications and consider all environmental factors;
- Use brainstorming in teams;
- Consider the human/machine interface;
- Consider usage mode changes;
- Try small-scale testing and theoretical analysis;
- Think through a worst-case what-if analysis.

In addition, additional sources of existing data may help with future predictions or determining such hazards. These hazard sources include:

- Accident reports/databases;
- Accident statistics;
- Near miss/dangerous occurrence reports;
- Reports from authorities or governmental bodies;
- Expert judgment.

A list of common accident databases:

eMARS – Major Accident Reporting System	This database is maintained by the Major Accident Hazards Bureau (MAHB) and is related to the EU Seveso II directive on major accidents involving highly hazardous substances.
FACTS – Failure and Accident Technical Information System	This database is operated by Unified Industrial & Harbor Fire Department in Rotterdam-Rozenburg
ARIA – Analyze, Recherche et Informations sur les Accidents	This database is operated by BARPI – a part of Ministère de l'Aménagement du Territoire et de l'Environnement)
PSID – Process Safety Incident Database	This database is operated by the Center for Chemical Process Safety CCPS)
NEDIES – Natural and Environmental Disaster Information Exchange System	This database is operated by the Joint Research Centre, Natural Risk Sector

ECCAIRS	European Co-ordination Centre for Aviation Incident Reporting Systems
Aviation Accident Database and Synopses	This database is operated by the National Transportation Safety Board, USA)
IRTAD	International Road Traffic and Accident Database (OECD)
WOAD – World Offshore Accident Database	This database is operated by Det Norske Veritas DNV
Accident statistics for fixed offshore unit son the UK Continental Shelf (UK HSE report)	
Accident statistics for mobile offshore units on the UK Continental Shelf (UK HSE report)	
Ship/platform collision incident database (UK HSE report)	
AirSafe.com	
Plane crash info	
Danger ahead – historic railway disasters	
BBC disaster news	

Below is a list of Accident Investigation Agencies in selected countries. Accident reports and accident statistics may be found on the web page of most of these agencies.

- U.S. National Transportation Safety Board (NTSB) is an independent U.S. Federal Government agency that carries out accident investigation of all types of transportation accidents.
- U.S. Chemical Safety Board (CSB) is responsible for investigating industrial chemical accidents in the U.S.
- UK Air Accident investigation Branch (AAIB) is part of the UK Department for Transport and is responsible for the investigation of civil aircraft accidents and serious incidents within the UK.
- UK Rail Accident Investigation Branch (RAIB) is responsible for investigating railway accidents within the UK.
- UK Marine Accident Investigation Branch (MAIB) is responsible for investigating marine accidents in the UK.
- "Statens Havarikommisjon for Transport" is responsible for investigation of transport accidents in Norway.
- Transportation Safety Board of Canada (TSB) is responsible for transport accident investigation in Canada.
- Australian Transport Safety Bureau (ATSB) is the agency responsible for investigating transport accidents in Australia

The risk related to a hazardous event is a function of the frequency of the event and the severity of its potential consequences. To determine the risk, we have to estimate both the frequency and severity of each hazardous event.

In addition, consequences must be considered. A hazardous event may lead to a wide range of consequences, ranging from negligible to catastrophic. In some applications, the severity of an average consequence of a hazardous event is assessed. In other applications, several possible consequences are considered, including the worst foreseeable consequence of the hazardous event.

The severity of an event may be classified in various ways. An example of such classification is presented below:

Rank	Severity Class	Description
4	Catastrophic	Failure results in major injury or death of personnel
3	Critical	Failure results in minor injury to personnel, exposure to harmful chemicals or radiation, fire, or the release of chemicals into the environment
2	Major	Failure results in a low level of exposure to personnel, or activates facility alarm system
1	Minor	Failure results in minor system damage but does not cause injury to personnel, allow any kind of exposure to operational or service personnel or allow any release of chemicals into the environment

Next comes estimating the frequency of an event. When estimating the frequency of an event, the relative consequences should also be considered. In some applications, we estimate the frequency of each hazardous event. To be used in a risk ranking, this frequency has to be related to the severity of an average consequence of each particular hazardous event. In other applications, we consider specific consequences of a hazardous event, such as the worst case. We must then estimate the frequency that the hazardous event produces a specific consequence. This may involve a combined assessment. For example, the frequency of the hazardous event, the probability that personnel are present, the probability that the personnel are not able to escape, and so on is such a combined event.

This, therefore, means that for each hazardous event, several consequences with associated frequencies may be presented. For example, consider a hazardous event where an operator falls from a work platform of low height.

In most cases, the consequence of such a fall will be a minor injury. This can be translated into low severity and high frequency. In a very rare case, the fall may result in a fatality. This would translate into high severity and very low frequency. Both consequences should be recorded.

The frequency of events may be classified into fairly broad categories. For example:

1	Very unlikely	Once per 1000 years or even more seldom
2	Remote	Once per 100 years
3	Occasional	Once per 10 years
4	Probable	Once per year
5	Frequent	Once per month or more often

The risk is then established as a combination of a given event/consequence and a severity of the same event/consequence. This enables ranking of the events/consequences in a risk matrix

Frequency/Consequence	**1 Very Unlikely**	**2 Remote**	**3 Occasional**	**4 Probable**	**5 Frequent**
Catastrophic					
Critical					
Major					
Minor					

	Acceptable – only ALARP actions considered.
	Acceptable – use ALARP principle and consider further investigations.
	Not acceptable – risk reducing measures required.

2.7. PROS AND CONS OF THE PHA METHOD

Each method of risk analysis has its own pros and cons. For the PHA method, the benefits include:

- Helping to ensure that the system is safe.
- Modifications are less expensive and easier to implement in the earlier stages of design.
- Decreases design time by reducing the number of surprises.

Cons include:

- Hazards must be foreseen by the analysts.
- The effects of interactions between hazards are not easily recognized.

2.8. FAULT TREE ANALYSIS (FTA)

Fault Tree Analysis (FTA) is a top-down approach to solving problems, starting with identifying the potential hazard and subsequently breaking it down into failure modes that may have caused the hazard. This method is sometimes referred to as "event tree analysis" and is used to identify the possible causes of a system failure. The method uses logic gates to relate potential failures to possible causes. A graphical illustration is used to depict the potential causes of failure in the form of a diagram.

The tree-like representation of root cause analysis provides a clear picture of each possible scenario considered. In this way, the factors contributing to an event (known as a failure) can be determined, as well as the probability of its occurrence. This method forms a deductive, top-down approach to determine the cause of a specific undesired event within a complex system and involves breaking down the root cause of a failure into its contributing factors.

2.9. HOW TO PERFORM A FAULT TREE ANALYSIS?

Performing a fault tree analysis is a complex process. However, the steps to carrying out a fault tree analysis are presented below:

1. **Define the Undesired Event:** Before running the analysis, the undesired event should be clearly defined. This event should be specific and measurable, such as a component failure or a system malfunction.
2. **Identify the Requirements for the Undesired Event to Occur/ Contributing Events and Factors:** Once the undesired event is defined, factors and events that might contribute to its occurrence should be identified. These contributing factors tend to fall into two broad categories: basic events and intermediate events.
   i. **Basic Events:** These are those events that cannot be further broken down into simpler events. They represent the most fundamental events in a fault tree and are the lowest level of events that can be analyzed. For example, a basic event in the case of a car accident might be that the driver loses control of the vehicle.

ii. **Intermediate Events:** These are located between the lower-level basic events and the top undesired event being analyzed. Intermediate events are caused by other events in the fault tree and, in turn, cause other events. These intermediate events represent higher-level events that can be analyzed further. For example, an intermediate event in a car accident might be that a tire blows out.

In both basic and intermediate events, both internal and external events should be considered, including component failures, human error, and environmental conditions. Subject matter experts may need to be consulted, and a review of historical data, incident reports, and maintenance records may be necessary.

3. **Decide on the Relationship between the Requirements and the Undesired Event:** Use the corresponding fault tree analysis symbols to construct the fault tree. This should be done using standard gate symbols and event symbols. A graphical representation of the relationships between the undesired event and its contributing factors should be constructed. The fault tree should be organized hierarchically, with the undesired event at the top and the contributing factors branching out below it. There are two main types of logic gates used in fault trees:

   i. **AND Gates:** These are used when all contributing events must occur simultaneously for the undesired event to occur.

   ii. **OR Gates:** These are used when any one of the input events is sufficient to cause the output event. In other words, the output event happens if at least one of the input events connected to the OR gate happens.

There are other gates that can be used, though they are less common. These include NOT gates, XOR gates, K/N gates, and INHIBIT gates:

a. **NOT Gates:** These are used to represent the inverse of an input event. If the input event does not occur, the output event will occur.

b. **XOR Gates (Exclusive OR Gates):** These are used when one of the input events must occur for the output event to happen. If none or more than one of the input events occurs, the output event will not happen.

c. **K/N Gates:** Also known as voting gates or threshold gates, these are used when a specific number of the input events (K) out of all the possible input events (N) must occur for the output event to happen. These gates are used to illustrate more complex relationships in a fault tree analysis.

d. **INHIBIT Gates:** These gates indicate that an output event will occur if both input events and a conditional event occur. This condition or restriction can apply to any gate.

By using these various available gates, a comprehensive fault tree that captures the complex interactions can be created. The tree can subsequently be refined as events continue to be broken down further.

4. **Gather Failure Data:** In this step, to quantify the risks associated with the undesired event, failure data should be gathered. This data can be sourced from historical records, industry databases, etc.

5. **Perform the Analysis:** Once the fault tree is constructed and failure data is gathered, an analysis can be performed. In this step, the probability of the undesired event occurring and identification of the most critical contributing factors should be calculated. This can be done using either qualitative or quantitative data analysis methods.

   i. **QualitativeAnalysis Techniques:** This focus on understanding the structure of the fault tree, the relationships between events, and the identification of critical paths, as well as any event that may have led to the undesired event. Qualitative analysis can help prioritize remedial actions and identify areas for further investigation.

   ii. **Quantitative Analysis:** This involves calculating the probability of the undesired event occurring, based on the failure probabilities of the basic events. This analysis can help inform risk management decisions and evaluate the effectiveness of proposed improvements.

6. **Interpret Results:** Once the analysis is done, interpreting the results is crucial, followed by the communication of relevant information to necessary stakeholders.

8. **Implement Improvements:** Based on the findings, the next step involves implementing preventive measures and improvements to decrease the likelihood of another event occurring.

2.10. PROS AND CONS OF FAULT TREE ANALYSIS (FTA)

As with all methods, Fault Tree Analysis (FTA) has its pros and cons. These are highlighted below:

- Pros:
  - A visual representation of contributing factors and events is presented, making it easier to understand complex interactions between system components.
  - This method allows for the calculation of the probability of a failure event occurring, enabling better risk management and decision-making.
  - The method focuses on one output event at a time, helping teams stay organized as they work systematically and methodically.
  - FTA accounts for human error, which can help identify whether issues are related to deviations from standard operating procedures.
  - FTA identifies which failures are likely to occur, which in turn helps prioritize which issues require urgent attention.
- Cons:
  - The accuracy and effectiveness of FTA rely heavily on:
    - o The expertise of the analysts.
    - o Their ability to identify relevant causes of failure.
    - o Their understanding of the complexities of the fault tree itself.
  - This method is best suited for smaller system analyzes. For large, complex systems, it may result in large, intricate fault trees, which can be time-consuming and challenging.
  - The availability of failure data may affect the precision of the calculated probabilities in a fault tree.
  - FTA allows you to examine only one top event at a time.

2.11. FAILURE MODE AND EFFECTS ANALYSIS (FMEA)

Failure Mode and Effects Analysis (FMEA) was developed in the 1950s and is one of the earliest structured risk methodologies available. FMEA is a methodology aimed at anticipating failure during the design stage by identifying all possible failures in a design or manufacturing process. This methodology forms a structured approach that allows for the discovery of potential failures within the design of a product or process.

Let's break down the methodology name in further detail to help understand the process. **Failure modes** are the ways in which a process can fail. **Effects** are the ways that these failures can lead to some related action, such as waste, defects, or harmful outcomes. Failure Mode and Effects Analysis is, therefore, designed to identify, prioritize, and limit these failure modes.

There are two broad categories of FMEA: **Design FMEA (DFMEA)** and **Process FMEA (PFMEA).**

2.11.1. Design FMEA

Design FMEA (DFMEA) focuses on the possibility of product malfunctions, reduced product life, and safety and regulatory concerns. These may arise from the following areas:

- Material properties;
- Geometry;
- Tolerances;
- Interfaces with other components and/or systems;
- Engineering Noise: environments, user profile, degradation, and systems interactions.

A template of a design FMEA worksheet is included below:

Potential Failure Mode and Effects Analysis (Design FMEA)
Core Team:
Revision:
FMEA Date:
Prepared by:

Item/ Function	Requirements	Potential Failure Mode	Potential Effects of Failure	SEV	Class	Potential Causes/ Mechanisms of Failure	Current Design Controls Prevention	Occurrence	DET	RPN	Recommended Actions	Responsibility and Target Completion Date	Actions Taken	SEV	OCC	DET	RPN

A template of a process FMEA worksheet is included below:

Potential Failure Mode and Effects Analysis (Process FMEA)
Core Team:
Revision:
FMEA Date:
Prepared by:

Item/ Function	Requirements	Potential Failure Mode	Potential Effects of Failure	SEV	Class	Potential Causes/ mechanisms of Failure	Current Process Controls Prevention	Occurrence	DET	RPN	Recommended Actions	Responsibility and Target Completion Date	Actions Taken	SEV	OCC	DET	RPN

2.11.2. Process FMEA (PFMEA)

Process FMEA (PFMEA) focuses on failures that impact product quality, reduce the reliability of the process, cause customer dissatisfaction, and present safety or environmental hazards. These failures may arise from the following areas:

- Human factors;
- Methods followed during processing;
- Materials used;
- Machines utilized;
- Measurement systems impacting acceptance;
- Environmental factors affecting process performance.

2.12. WHY IS THE FMEA METHOD PREFERRED?

The FMEA method is generally preferred because it is one of the tools used to discover failures at the earliest possible point in product or process design. Identifying failure early on offers several benefits, including:

- Allowing multiple choices for mitigating risk.
- Providing higher capability for making verification and validation changes.
- Facilitating collaboration between product and process design.
- Improving design for manufacturing and assembly.
- Offering lower-cost solutions.

The FMEA method is also preferred because it effectively identifies and corrects process failures early on, preventing poor performance.

2.13. WHEN SHOULD YOU PERFORM AN FMEA?

There are several key times when it makes sense to perform a Failure Mode and Effects Analysis, including:

- When a new product, process, or service is being designed.
- When an existing process is going to be performed in a different way.
- When quality improvement goals for a specific process exist.
- When understanding and improving the failures of a process is needed.

Additionally, an FMEA should be occasionally performed throughout the lifetime of a process or product.

2.14. HOW TO PERFORM AN FMEA?

FMEA is performed in seven steps, which are detailed below:

1. **FMEA Pre-Work and Assembly of the FMEA Team:** Pre-work involves collecting and creating key documents. Investigating past failures and gathering preparatory documents should begin at the onset. Such documents may include:
    - Failure mode avoidance (FMA) Past Failures;
    - Boundary/block diagram (for DFMEA);
    - Parameter diagram (for DFMEA);
    - Process flow diagram (for PFMEA);
    - Characteristics matrix (for PFMEA).

A pre-work checklist is recommended for an efficient FMEA event. Items on the checklist may include:

- Requirements to be included.
- Design and/or process assumptions.
- Preliminary bill of materials/components.
- Known causes from surrogate products.
- Potential causes from interfaces.
- Potential causes from design choices.
- Potential causes from noise and environments.
- Family or baseline FMEA (historical FMEA).
- Past test and control methods used on similar products.

2. **Path 1 Development (Requirements through Severity Ranking):** Path 1 involves inserting the functions, failure modes, effects of failure, and severity rankings. The pre-work documents assist by capturing previous information to populate the first few columns of the FMEA.

- Each function must have an associated measurable. Functions may include:
    - o Wants, needs, and desires translated.
    - o Specifications of a design.
    - o Government regulations.
    - o Program-specific requirements.
    - o Characteristics of the product to be analyzed.
    - o Desired process outputs.

- Failure modes are written as anti-functions or anti-requirements. This is done in five potential ways:
    - o Full function failure.
    - o Partial/degraded function failure.
    - o Intermittent function failure.
    - o Over-function failure.
    - o Unintended function failure.
- Effects are the results of failure, where each individual effect is given a severity ranking. Actions should be considered if the severity is 9 or 10.

3. **Path 2 Development (Potential Causes and Prevention Controls through Occurrence Ranking):** Causes are selected from design/ process inputs or past failures and placed in the cause column when applicable to a specific failure mode. Information to be completed includes:
    - Potential causes/mechanisms of failure.
    - Current prevention controls.
    - Occurrence rankings for each cause.
    - Classification of special characteristics, if indicated.
    - Actions are developed to address high-risk severity and occurrence combinations, defined in the quality-one criticality matrix.
4. **Path 3 Development (Testing and Detection Controls through Detection Ranking):** Path 3 involves adding detection controls that verify the design meets the requirements (for Design FMEA) or that cause/failure modes, if undetected, may reach a customer (for Process FMEA). The columns completed in Path 3 are:
    - Detection controls.
    - Detection ranking.

Actions are determined to improve controls if they are insufficient to address the risks identified in Paths 1 and 2. Recommended actions should address weaknesses in the testing and/or control strategy.

5. **Action Priority and Assignment:** The actions determined in Paths 1, 2, or 3 are assigned a Risk Priority Number (RPN) for follow-up. The RPN is calculated by multiplying the severity, occurrence, and detection rankings for each potential failure/effect, cause, and control combination.

Actions should not be based solely on an RPN threshold value, as this practice can lead to poor team behavior. Instead:

- Review recommended actions and assign RPN for follow-up.
- Assign actions to appropriate personnel.
- Assign action due dates.

6. **Actions Taken/Design Review:** FMEA actions are closed when countermeasures have been taken and are successful in reducing risk. An FMEA that does not uncover risk is considered weak and non-value-added. If no improvement is produced, the time spent on analysis is wasted.
7. **Re-Ranking RPN and Closure:** After successful confirmation of risk mitigation actions, the core team or team leader will re-rank the severity, occurrence, or detection values. The new rankings will be multiplied to attain the new RPN. The original RPN is compared to the revised RPN, and relative improvement in the design or process is confirmed. The columns completed in this step include:
   - Re-ranked severity.
   - Re-ranked occurrence.
   - Re-ranked detection.
   - Re-ranked RPN.
   - Generation of new actions, repeating Step 5, until risk has been mitigated.
   - Comparison of initial RPN and revised RPN.

2.15. HAZARD AND OPERABILITY ANALYSIS (HAZOP)

Hazard and Operability Analysis (HAZOP) is a systematic approach used to identify potential problems during the review of design safety or while revisiting existing processes and operations. It is commonly applied in industries such as chemical, pharmaceutical, oil and gas, and nuclear. This technique, also known as a HAZOP study or HAZOP analysis, is based on the theory that risk events are caused by deviations from design or operating principles. Identifying these deviations is facilitated by using sets of "guide words" that offer a systematic list of deviation perspectives. This approach is unique because it helps stimulate the imagination of team members when exploring potential deviations.

HAZOP is often described as a brainstorming technique or a qualitative risk assessment tool, and an inductive risk assessment tool, which ultimately means that a bottom-up approach is used. The effectiveness of HAZOP relies on the ability of subject matter experts (SMEs) to predict deviations based on past experiences and their general expertise.

When utilizing the HAZOP methodology, the following definitions can be useful:

Hazard	Potential source of harm. Deviations from design or operational intent which may constitute or produce a hazard. A single hazard could potentially lead to multiple forms of harm.
Harm	Physical injury or damage to the health of people or damage to property or the environment. Harm is the consequence of a hazard occurring and may take many forms: patient or user safety, employee safety, business risks, regulatory risks, environmental risks, etc.
Risk	Combination of probability of occurrence of harm and the severity of that harm.

HAZOP is best suited for assessing hazards in facilities, equipment, and processes, and it is capable of evaluating systems from multiple perspectives. These perspectives include design, physical and operational environments, as well as operational and procedural controls.

A simple HAZOP methodology is presented below:

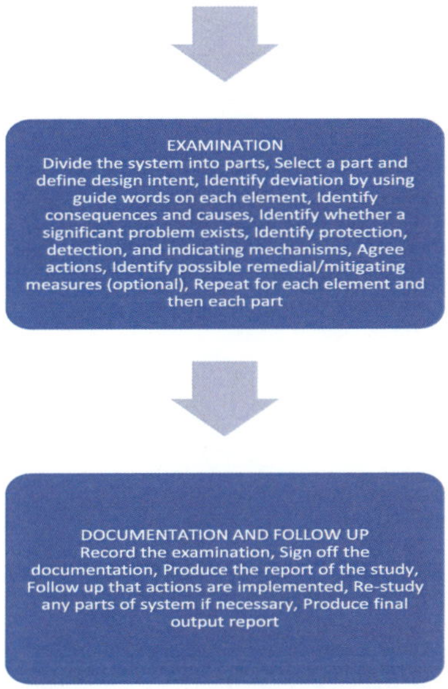

This is explained in greater detail below:

1. **Definition Phase:** The **definition phase** begins with a preliminary identification of the risk assessment team members. The team should consist of specialists across various disciplines who possess the appropriate skills, experience, and good intuition and judgment. Subject matter experts should be carefully selected to include those with broad and current knowledge of system deviations. During the Definition Phase, the risk assessment team must identify the assessment scope carefully to focus efforts effectively. This includes defining boundaries, key interfaces, and key assumptions under which the assessment will be performed.
2. **Preparation Phase:** The **preparation phase** typically includes the following activities:
    - **Identification and location of supporting data and information**
    - **Identification of the users of the study outputs**
    - **Project management preparations** – This includes the scheduling of meetings, taking of minutes, etc.
    - **Consensus on template format for recording study outputs**
    - **Consensus on HAZOP guide words** to be used during the study

Such guide words are listed in the box below:

- No
- Not
- More
- Less
- As well as
- Part of
- Reverse (of intent)
- Other than
- Early
- Late
- Before
- After

The HAZOP guide words work by providing a systematic and consistent means of brainstorming potential deviations to operations. A good example of utilization of these guidewords is provided in the table below on how different HAZOP guide words might be used to brainstorm deviations:

- No – no chemical added
- Not – the chemical was not used
- More – too much volume of the chemical was added
- Less – too little chemical volume was added/chemical concentration is too low
- Part of – critical chemical components were omitted
- Reverse (of intent) – chemical is contaminated with…
- Other than – wrong chemical used
- Early – chemical is added too early
- Late – chemical is added to late
- Before – chemical is added before another component
- After – chemical is added after another component

3. **Examination Phase:** Once the guide words have been selected, the **Examination Phase** can begin. This phase starts with the identification of all elements of the system or process to be examined. These elements include:

   - **Physical systems**, which may be broken down into smaller parts as necessary.

- **Processes**, which may be divided into discrete steps or phases.
- **Similar parts or steps**, which may be grouped together to facilitate assessment.

Once the elements have been identified, the HAZOP guide words can be applied to each of the elements. It is important to note that not all combinations of guide words and elements will yield sensible or credible deviations. Additionally, all reasonable use and misuse conditions expected by the user should be identified and challenged to determine if they are "credible" and whether they should be assessed further.

Figure 2.1 graphically illustrates the HAZOP examination phase process flow.

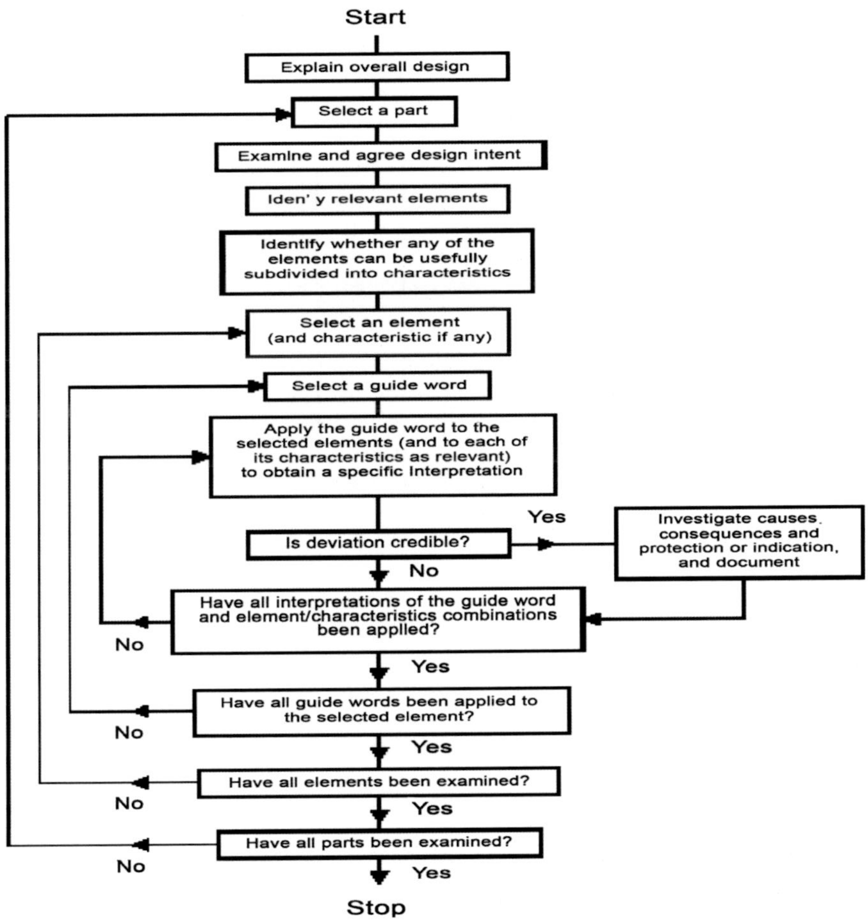

Figure 2.1. *The HAZOP examination phase process flow.*

Generally, HAZOP analyzes are documented using standard templates, as outlined in the **IEC 61882** standard. These templates may be modified to include regulatory requirements, company documentation policies, traceability/audit readiness, and other necessary documentation elements.

An example of a template is presented below:

Number	**Guide Word**	**Element**	**Deviation**	**Possible Causes**	**Consequences**	**Safeguards**	**Comments**	**Actions Required**	**Actions Assigned to**
Assign each entry a unique tracking number	Insert deviation guide word used	Describe the deviation	Describe how the deviation may occur	Describe how the deviation may occur	Describe what may happen if the deviation occurs	List controls (preventive or reactive) that reduce deviation likelihood or severity	Capture key rationale, assumptions, data, etc.	Identify any hazard mitigation or control actions required	Record who is responsible for actions
1.									
2.									
3.									
4.									
5.									
6.									
7.									
8.									
9.									
10.									
11.									
12.									

2.16. HAZARD ANALYSIS AND CRITICAL CONTROL POINT (HACCP)

Hazard Analysis and Critical Control Point (HACCP) is a systematic approach to food safety management that helps identify and prevent potential hazards in the food industry, specifically in the production and processing sectors. It is a preventive system rather than a reactive one, designed to ensure that all necessary steps are taken to produce food products that are compliant and safe.

HACCP is based on **seven principles**, which form the foundation of a HACCP plan. The first two principles establish the foundation, while the remaining five provide the structure for implementing the actions of the HACCP plan within a processing plant.

2.16.1. Principle 1: Conduct a Hazard Analysis

This step involves listing the steps in the process and identifying where significant hazards are likely to occur. The HACCP team focuses on hazards that can be prevented, eliminated, or controlled as identified by the HACCP plan. A justification for including or excluding each hazard should also be documented, along with possible control measures.

2.16.2. Principle 2: Determine Critical Control Points (CCPs)

A **Critical Control Point (CCP)** is a point, step, or procedure where control can be applied to prevent, eliminate, or reduce a food safety hazard to an acceptable level. A **CCP decision tree** is often used to identify these points in the process. It is important to note that a CCP may control more than one food safety hazard, and in some cases, multiple CCPs may be needed to control a single hazard.

2.16.3. Principle 3: Establish Critical Limits

A **critical limit** is the maximum and/or minimum value to which a biological, chemical, or physical parameter must be controlled at a CCP to prevent, eliminate, or reduce a food safety hazard to an acceptable level. Critical limits are usually measures such as time, temperature, water activity, pH, weight, or other parameters based on scientific literature and/or regulatory standards.

2.16.4. Principle 4: Establish Monitoring Procedures

Monitoring procedures are used to measure the critical limit at each CCP. These procedures should detail:

- How the measurement will be taken.
- When the measurement will be taken.
- Who is responsible for the measurement.
- The frequency of measurement during production.

2.16.5. Principle 5: Establish Corrective Actions

Corrective actions are the procedures to follow when there is a deviation from a critical limit. The HACCP team identifies steps to prevent potentially hazardous food from entering the food chain and to correct the process.

2.16.6. Principle 6: Establish Verification Procedures

Verification activities, apart from monitoring, ensure the validity of the HACCP plan and confirm that the system is operating according to the plan. Verification procedures may include auditing CCPs, reviewing records, inspecting shipments, calibrating equipment, and conducting product testing.

2.16.7. Principle 7: Establish Record-Keeping and Documentation Procedures

Record-keeping is a crucial aspect of the HACCP plan. Records should document that food was produced safely and include information about:

- The HACCP plan.
- The HACCP team.
- Product description.
- Flow diagrams.
- Hazard analysis.
- CCPs identified.
- Critical limits.
- Monitoring systems.
- Corrective actions.
- Record-keeping procedures.
- Verification procedures.

CHAPTER 3

UNDERSTANDING THE RISK MANAGEMENT PROCESS

Contents

3.1. OVERVIEW OF THE RISK MANAGEMENT PROCESS

Risk management involves the identification, understanding, control, and prevention of failures that could result in hazards when people use medical devices. It is the responsibility of manufacturers to identify possible hazards associated with the design of the device in both normal and fault conditions, as well as after the design and manufacture have been completed.

3.2. IMPORTANT DEFINITIONS IN THE RISK MANAGEMENT PROCESS

A selection of important definitions related to risk management is provided below. These terms are often used when discussing the risk management of medical devices.

Term	**Definition**
Benefit	Positive impact or desirable outcome of the use of a medical device on the health of an individual, or a positive impact on patient management or public health. These benefits can include positive impact on clinical outcome, the patient's quality of life, outcomes related to diagnosis, positive impact from diagnostic devices on clinical outcomes, or positive impact on public health.
Harm	Injury or damage to the health of people, or damage to property or the environment.
Hazard	Potential source of harm
Hazardous situation	Circumstance in which people, property or the environment is/are exposed to one or more hazards.
Intended use	Use for which a product, process or service is intended according to the specifications, instructions and information provided by the manufacturer. Note: The intended medical indication, patient population, part of the body or type of tissue interacted with, user profile, use environment and operating principle are typical elements of the intended use
Life cycle	Series of all phases in the life of a *medical device* from the initial conception to final decommissioning and disposal.

Manufacturer	Natural or legal person with responsibility for the design and/or manufacture of a *medical device* with the intention of making the *medical device* available for use, under his name, whether or not such a *medical device* is designed and/or manufactured by that person himself or on his behalf by another person(s).
Medical device	Instrument, apparatus, implement, machine, appliance, implant, reagent for in vitro use, software, material or other similar or related article, intended by the *manufacturer* to be used, alone or in combination, for human beings, for one or more of the specific medical purpose(s) of: · Diagnosis, prevention, monitoring, treatment or alleviation of disease, · Diagnosis, monitoring, treatment, alleviation of or compensation for an injury, · Investigation, replacement, modification, or support of the anatomy or of a physiological process, · Supporting or sustaining life, · Control of conception, · Disinfection of *medical devices*, · Providing information by means of in vitro examination of specimens derived from the human body, and which does not achieve its primary intended action by pharmacological, immunological or metabolic means, in or on the human body, but which may be assisted in its function by such means.
Post-production	Part of the *life cycle* of the *medical device* after the design has been completed and the *medical device* has been manufactured.
Reasonably foreseeable misuse	Use of a product or system in a way not intended by the manufacturer, but which can result from readily predictable human behavior. *Note:* Readily predictable human behavior includes the behavior of all types of users, e.g., lay and professional users. Reasonably foreseeable misuse can be intentional or unintentional.
Residual risk	Risk remaining after risk control measures have been implemented.

Risk	Combination of the probability of occurrence of harm and the severity of that harm.
Risk control	Process in which decisions are made and measures implemented by which risks are reduced to, or maintained within, specified levels.
Safety	Freedom from unacceptable risk
Risk analysis	This is the process of looking at a design or system and identifying possible hazards that could cause harm to people, property, or even the environment. Think of this as sitting around a table with your team and coming up with any possible, often seemingly silly, hazard that could arise.
Risk estimation	This is the process used to assign a numerical value to the laundry list of possibilities you came up with during the analysis. The estimation process considers the probability of occurrence and the severity of the harm.
Risk evaluation	This is a process of taking the risk analysis and comparing it to a set of predefined risk acceptance criteria. Usually this results in a risk evaluation summary document.
Risk assessment	This is a more comprehensive document that contains both the risk analysis and the risk evaluation.
Risk control	This is the step where the full risk assessment is reviewed and steps are taken to reduce risk to an acceptable level.
Risk management	The overall process of all things risk. Risk management is the umbrella term that refers to all the sub-processes and components.
Risk management file	The records for all things risk management including analysis, estimation, acceptable risk ranges, mitigations, etc.
Safety	Freedom from unacceptable risk.
Severity	Measure of the possible consequences of a hazard.
Use error	User action or lack of user action while using the medical device that leads to a different result than that intended by the manufacturer or expected by the user.

CHAPTER 4

ISO AND RISK STANDARDS

Contents

4.1. OVERVIEW OF ISO AND RISK STANDARDS

Medical devices are subject to strict controls and regulations to ensure safety and efficiency are met at all times. To meet these safety and efficacy requirements, various standards have been developed by the International Standards Organization (ISO) and are currently in use. Understanding the various standard systems available, together with the standards development process and their use in conformity assessments, is imperative to understanding how regulatory and risk controls are enforced.

Various regulatory agencies and standard organizations have collaborated to establish acceptable standards and standard-setting activities for medical devices. These standard-setting activities include the development of performance characteristics, characterization and testing methodologies, manufacturing practices, scientific protocols, compliance criteria, ingredient specifications, labeling, and/or other technical criteria.

4.2. WHAT ARE STANDARDS?

The International Standards Organization has put forward a formal definition of a standard that should be adopted where medical devices are concerned. This definition is as follows:

- "Standards are documented agreements containing technical specifications or other precise criteria to be used consistently as rules, guidelines, or definitions of characteristics, to ensure that materials, products, processes, and services are fit for their purpose."
- International standards form the building blocks for harmonized regulatory processes to assure the safety, quality, and performance of medical devices. Regulatory Authorities and various Standards Organizations should encourage and support the development of international standards for medical devices to demonstrate compliance with critical safety and performance requirements.

4.3. TYPES OF STANDARD SPECIFICATIONS

Standards can establish a wide range of specifications for products, processes, and services. The various types of specifications in standards are outlined below:

1. **Prescriptive Specifications:** These specifications necessitate product characteristics such as device dimensions, biomaterials, and test and/or calibration procedures. Prescriptive specifications also determine definitions of terms and terminologies.

2. **Design Specifications:** These specifications set out the specific design or technical characteristics of a product.
3. **Performance Specifications:** These specifications ensure that a product meets a prescribed test. This includes strength requirements, measurement accuracy, battery capacity, and so on.
4. **Management Specifications:** These specifications set out requirements for the processes and procedures companies put in place, such as quality systems for manufacturing or environmental management systems.

In recent times, the development and application of what are known as "generic management system standards" have become more apparent. The term "generic" ideally refers to standards' requirements that can be applied to any organization regardless of the product it makes or the service it delivers, while the phrase "management system" refers to what the organization does to manage its processes. Terms such as "outcome-oriented" standards, "objective" standards, "function-focused" standards, and "result-oriented" standards are also used. In simpler terms, these kinds of standards refer to those standards that specify the objectives or ends to be achieved while leaving the methods to the implementers.

4.4. THE PURPOSE OF STANDARDS

International standards serve a range of different purposes. These include:

- Provide reference criteria that a product, process, or service must meet;
- Provide information that enhances the safety, reliability, and performance of products, processes, and services;
- Assure consumers about the reliability or other characteristics of goods or services provided in the marketplace;
- Give consumers more choice by allowing the product of one company to be substituted for, or combined with, the product of another company.

The increasing globalization of medical devices makes the need and subsequent benefits provided from such standards extremely important, especially on an international level where manufacturing, trade, and communications are concerned. As more and more medical devices are used globally, the need for global communication, and as a result, international standardization has increased in an attempt to ensure safety, performance, and consistent quality all around.

4.5. UNDERSTANDING THE DIFFERENT TYPES OF STANDARDS AVAILABLE

4.5.1. Voluntary vs. Mandatory Standards

Most standards are voluntary; however, a standard may be mandated or delegated by a company, professional society, industry, government, or trade agreement. If a standard becomes mandated, it may then be called a regulation and may, or may not, have a legal basis. When a standard is mandated by a government or an international trade agreement, it normally becomes legally obligatory. This is often based on regulations or a law established by the government or the contracts between such international bodies.

4.6. DEVELOPMENT PROCESS OF STANDARDS

A range of steps is used by development organizations when creating standards. Essentially, experts form a technical committee that is responsible for a particular subject area, and this team then begins the development process. It starts with the creation of a draft that meets a specific market need, which is then shared for comment and discussion. This is followed by some form of voting process, and once consensus is achieved, the draft is on its way to becoming a standard. If, however, agreement is not reached, the draft will be further modified and re-voted on until consensus is achieved.

4.7. KEY PRINCIPLES IN THE DEVELOPMENT OF ISO STANDARDS (AS DEFINED BY ISO)

1. **The Standards Should Respond to a Need in the Market**: Ideally, a standards development organization should respond to a request from the market, industry, or consumer groups when deciding to develop a standard. Usually, what happens is an industry sector or group communicates the need for a standard to its national member, who then contacts a standards development organization like ISO.
2. **The Standards Should Be Based on Global Expert Opinion**: International standards should be developed by groups of experts from all over the world. These experts should ideally be part of larger groups called technical committees, who will negotiate all aspects of the standard, including the scope, definitions, and overall content.
3. **The Standards Should Be Developed Through a Multi-Stakeholder Process:** The technical committees are made up of

experts from the relevant industry concerned, as well as consumer associations, academia, NGOs, and government.

4. **Standards Should Be Based on Consensus**: A consensus-based approach should always be adopted when developing international standards, allowing the comments and opinions of all stakeholders to be taken into consideration.

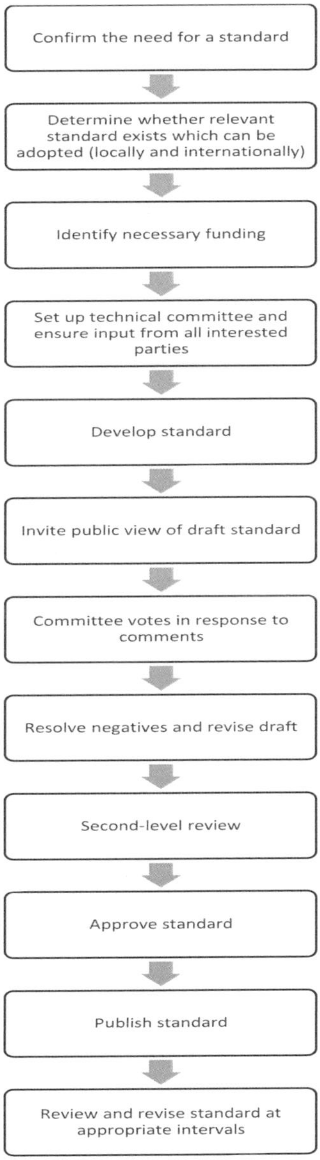

4.8. ATTRIBUTES OF GOOD STANDARDS

- Their development has been overseen by a recognized body, ensuring that the process is transparent and not biased.
- The development process has been open to input from all interested parties, and the resulting document is based on a consensus among relevant stakeholders.
- It is based on consolidated results of science, technology, and experience, ultimately aimed at the promotion of optimum community benefits.
- The standards do not hinder innovation and must be periodically reviewed to remain in tune with technological advances.

4.9. CONFORMITY ASSESSMENT WITH STANDARDS

There are four common methods for assessing conformity to a particular standard:

- A product's conformity to a standard is assessed by direct testing.
- A process can be assessed by an audit in which certification organizations or regulatory authorities attest that products or processes conform to a standard. This is usually done by authorizing the display of the certification organization or regulatory authority's (RA) certification mark.
- The conformity to a management standard, known as management systems registration, involves formal audit procedures and the registration of certification for a company.
- **Accreditation**, which is used by an authoritative body to give formal recognition that an organization or a person is competent to carry out a specific task. A common example is the notified bodies (NBs) in Europe, which carry out conformity assessments of medical devices.

4.10. NATIONAL AND INTERNATIONAL STANDARDS SYSTEMS

While a country can have many voluntary standard bodies, ideally, there is one official national organization that coordinates and accredits standard development and standard development bodies in the country. This official national organization would have the authority to endorse a document as a national standard while also representing the country in various international standards organizations. These organizations differ from country to country, for example, in the United States, it is the American National Standards Institute (ANSI), and in Canada, it is the Standards Council of Canada (SCC).

The presence of a standards system not only helps medical device administration but is also important for other industrial and economic developments. Having a standardized infrastructure has now become a basic requirement for the successful implementation of policies that ideally aim to improve productivity, market competitiveness, and export capability.

The three major international standardization organizations are:

1. The International Organization for Standardization (ISO);
2. The International Electrotechnical Commission (IEC);
3. The International Telecommunication Union (ITU).

Each of these international standardization organizations covers different areas. ITU covers telecommunications, IEC covers electrical and electronic engineering, and ISO covers anything else. There are other organizations known to produce international standardization documents that are subsequently adopted by ISO/IEC/ITU. These standards are then adopted as if they were developed in accordance with international consensus criteria.

Links to these three major international standardization organizations are given below:

- The International Organization for Standardization (ISO): www.iso.ch.
- The International Electrotechnical Commission (IEC): www.IEC.ch.
- The International Telecommunication Union (ITU): www.itu.int.

4.11. IDENTIFICATION OF STANDARDS

Standards are generally named by an alphabetical prefix and a number. The letters indicate the international body that has approved them, such as ISO, IEC, ANSI, CAN, EN, and DIN, to name a few; while the numbers identify the specific standard and the year in which it was finalized. In certain instances, the standard reference code will give an indication of adoption where standards are equivalent. For example:

- ANSI/AAMI/ISO 15223:2000 means the international standard ISO 15223 (established in 2000) adopted by the Association for the Advancement of Medical Instrumentation in the United States, which, in turn, is designated by the American National Standards Institute (ANSI) as an American national standard.

- UNI EN ISO 9001 indicates an Italian national standard (UNI), which is an adoption of a European standard (EN), which is itself an adoption of the International Standard ISO 9001.

4.12. THE USE OF STANDARDS IN MEDICAL DEVICE REGULATIONS (MDRS)

The use of voluntary standards came about due to the realization that, while regulations generally address essential safety and performance principles, manufacturers and users still need to know detailed specifications pertaining to specific products. Providing these specifications and detailed requirements, however, presents a mammoth task for regulatory authorities. The large number of voluntary standards available or currently being developed offered an alternative way to provide such precise specifications.

The use of voluntary standards has many advantages, including:

- These standards are normally developed by experts who have access to a range of resources in the professional and industrial communities.
- Through existing resources, the government can overcome its own limited resources when providing product-specific technical requirements and characteristics.
- Conformity to standards can be assessed by an accredited third party, such as a notified body in Europe.
- The use of international standards facilitates harmonized regulatory processes, thereby improving global access to new technology.
- It is much easier to update standards than to change regulations as technology advances and changes.
- Manufacturers have the flexibility to choose appropriate standards or other means to demonstrate compliance with regulatory requirements.

4.13. REGULATIONS AND STANDARDS FOR RISK MANAGEMENT OF MEDICAL DEVICES

To date, almost every medical device regulatory authority has included risk as a critical component in medical device regulation. The most well-known regulators, such as the US FDA, Health Canada, EU Competent Authorities, Australian TGA, and Japanese MHLW, all require a defined risk management process as well as risk management documentation before a device is approved and placed on the market.

It is also important to note that, apart from the key standard of ISO 14971, there are other medical device standards that require risk management as well. These standards incorporate risk requirements into the overall standard. Such standards are presented in the table below:

IEC 60601	Standards for medical device electric equipment.
IEC 62366	Standards that cover usability engineering.
ISO 10993	Standards for Biocompatibility.
ISO 13485	Medical Devices – Quality Management System – Requirements for regulatory purposes.

4.14. UNDERSTANDING IEC 60601: STANDARDS FOR MEDICAL DEVICE ELECTRICAL EQUIPMENT

IEC 60601 is a series of international standards published by the International Electrotechnical Commission (IEC). These standards specify safety and performance requirements for medical electrical equipment and consist of multiple parts, each focusing on different aspects of medical electrical equipment. These include:

Standard	**Title**	**Description**
IEC 60601-1-2 Medical electrical equipment – Parts 1–2	General requirements for basic safety and essential performance – Collateral Standard: Electromagnetic disturbances – Requirements and tests	This standard applies to the basic safety and essential performance of medical equipment and medical equipment systems in the presence of electromagnetic disturbances and to electromagnetic disturbances emitted by medical equipment and related systems.
IEC 60601-1-3 Medical electrical equipment – Parts 1–3	General requirements for basic safety and essential performance – Collateral Standard: Radiation protection in diagnostic X-ray equipment.	This standard applies to X-ray equipment and to subassemblies of such equipment where radiological images of a human patient are used for diagnosis, planning or guidance of medical procedures. The standard focuses on establishing general requirements for protection against X-radiation in X-ray equipment.

IEC 60601-1-6 Medical electrical equipment – Parts 1–6	General requirements for basic safety and essential performance – Collateral standard: Usability	The IEC 60601-1-6 standard specifies the process for any medical device manufacturer to analyze, specify, design, verify and validate usability in relation to basic safety and essential performance of medical electrical equipment. The standard offers a framework for manufacturers to ensure their medical electrical equipment (ME equipment) achieves high standards of product safety and effectiveness through focused usability testing. The standard provides details on how to analyze, plan, develop and evaluate the usability of ME equipment.
IEC 60601-1-8 Medical electrical equipment – Parts 1–8	General requirements for basic safety and essential performance – Collateral Standard: General requirements, tests and guidance for alarm systems in medical electrical equipment and medical electrical systems	This standard specifically addresses medical alarm systems. The standard includes instructions on when alarms should be triggered and how their priority is classified as well as basic safety and essential performance requirements. in addition, tests for alarm systems in medical electrical equipment and medical electrical systems is provided.
IEC 60601-1-9 Medical electrical equipment – Parts 1–9	General requirements for basic safety and essential performance – Collateral Standard: Requirements for environmentally conscious design	This standard focuses on improving the environmental impact for the entire range of medical electrical equipment across all stages of the product life cycle from product specification to design, manufacturing, sales, logistics, installation and up to end of life. The purpose of the standard is to protect the environment and people from hazardous substances, reduce consumption of natural resources and minimize environmental impact from waste.

IEC 60601-1-10 Medical electrical equipment – Parts 1–10	General requirements for basic safety and essential performance – Collateral Standard: Requirements for the development of physiologic closed-loop controllers.	This standard focuses on the requirements for the development of a physiological closed loop controller (PCLC) as part of a physiological closed loop controller system (PCLCS) in medical electrical equipment and medical electrical systems.
IEC 60601-1-11 Medical electrical equipment – Parts 1–11	General requirements for basic safety and essential performance – Collateral Standard: Requirements for medical electrical equipment and medical electrical systems used in the home healthcare environment.	This standard focuses on basic safety and essential performance of medical electrical equipment and medical electrical systems which are intended by the manufacturer for use in the home healthcare environment. This is irrespective of whether the medical electrical equipment or medical electrical system is intended for use by a lay operator or by trained healthcare personnel. The standard specifies general requirements in addition to IEC 60601-1:2005. In addition, it serves as the basis for particular standards.
IEC 60601-1-12 Medical electrical equipment – Parts 1–12	General requirements for basic safety and essential performance – Collateral Standard: Requirements for medical electrical equipment and medical electrical systems intended for use in the emergency medical services environment.	This standard applies to the basic safety and essential performance of medical electrical equipment and medical electrical systems which are intended for use by manufacturers in an EMS environment. The standard does not apply to equipment and systems intended for use solely in a home healthcare environment or professional healthcare facilities.

IEC 60601-2-1 Medical electrical equipment – Parts 2–1	Particular requirements for the basic safety and essential performance of electron accelerators in the range 1 MeV to 50 MeV.	This standard focuses on the requirements for basic safety and performance of electron accelerator devices in the range of 1 to 50 million electron volts used in the treatment of patients.
IEC 60601-2-2 Medical electrical equipment – Parts 2–2:	Particular requirements for the basic safety and essential performance of high frequency surgical equipment and high frequency surgical accessories.	This standard applies to the basic safety and essential performance of high frequency surgical equipment and high frequency surgical accessories.
IEC 60601-2-3 Medical electrical equipment – Parts 2–3:	Particular requirements for the basic safety and essential performance of short-wave therapy equipment.	This standard focuses on the safety of short-wave therapy equipment having a rated output power not exceeding 500 W.
IEC 60601-2-4 Medical electrical equipment – Parts 2–4	Particular requirements for the basic safety and essential performance of cardiac defibrillators	This standard focuses on particular requirements for basic safety and essential performance requirements for cardiac defibrillators.
IEC 60601-2-5 Medical electrical equipment – Parts 2–5	Particular requirements for the basic safety and essential performance of ultrasonic physiotherapy equipment.	This standard focuses on the requirements for safety and essential performance of ultrasonic physiotherapy equipment which is used in medical practice. This standard does not apply to equipment in which a tool is driven by ultrasound (i.e., for example, equipment used in surgery or dentistry) or equipment in which focused ultrasound pulse waves are used to destroy conglomerates such as stones in the kidney or the bladder.

IEC 60601-2-6 Medical electrical equipment – Parts 2–6	Particular requirements for the basic safety and essential performance of microwave therapy equipment.	This standard specified the requirements for the safety of microwave therapy equipment which is used in medical practice. The standard focuses on establishing the basic standard of safety and essential performance requirements for microwave therapy.
IEC 60601-2-8 Medical electrical equipment – Parts 2–8	Particular requirements for basic safety and essential performance of therapeutic X-ray equipment operating in the range 10 kV to 1 MV.	This standard focuses on establishing the particular requirements for basic safety and essential performance of therapeutic X-ray equipment with nominal voltages in the range of 10 kV to 1mV. The standard includes the requirements for accuracy and reproducibility of performance to the extent that these are related to radiation quality and the quantity of ionizing radiation produced.
IEC 60601-2-10 Medical electrical equipment – Parts 2–10	Particular requirements for the basic safety and essential performance of nerve and muscle stimulators.	This standard specifies the requirements for the safety and essential performance of nerve and muscle stimulators, for use in the practice of physical medicine. This includes transcutaneous electrical nerve stimulators (TENS) and electrical muscle stimulators (EMS).
IEC 60601-2-11 Medical electrical equipment – Parts 2–11	Particular requirements for the basic safety and essential performance of gamma beam therapy equipment.	This standard specifies the requirements for the safety of gamma beam therapy equipment which is intended for radiotherapy in human medical practice and includes equipment in which the selection and display of operating parameters can be controlled by a programmable electronic system. The standard applies to equipment which is intended to deliver gamma radiation beams at normal treatment distances greater than 5 cm using sealed radioactive source.

		The standard establishes requirements to ensure the radiation safety and enhance the electrical and mechanical safety of gamma beam therapy equipment used in human medical practice and specifies tests for demonstrating compliance with those requirements.
IEC 60601-2-12 Medical electrical equipment – Parts 2–12:	Particular requirements for the basic safety and essential performance of critical care ventilators.	This standard applies to the basic safety and essential performance of critical care ventilators in combination with its accessories. It focuses on ventilators which is intended for use in an environment that provides specialized care for patients whose conditions can be life threatening and who require comprehensive care and constant monitoring in a healthcare facility. This document is also applicable to those accessories which are intended by the manufacturer to be connected to a ventilator breathing system, or to a ventilator, where characteristics of the accessories can affect the basic safety or essential performance of the ventilator. This standard is not applicable to ME equipment or an ME system operating in a ventilator operational mode which is solely intended for patients who are not dependent on artificial ventilation. The standard is not applicable to: • ME equipment that is intended solely to augment the ventilation of spontaneously breathing patients in a healthcare facility The standard does not specify requirements for: • Ventilators or accessories intended for anaesthetic applications as per ISO 80601-2-13 • Ventilators or accessories intended for emergency medical services environment as per ISO 80601-2-84

		• Future replacement of ISO 10651-3 • Ventilators or accessories intended for ventilator-dependent patients in the home healthcare environment as per ISO 80601-2-72:2015 • Ventilators or accessories intended for home-care ventilatory support devices obstructive sleep apnoea therapy ME equipment • Continuous positive airway pressure ME equipment • High frequency jet ventilators and high frequency oscillatory ventilators
IEC 60601-2-13 Medical electrical equipment – Parts 2–13:	Particular requirements for basic safety and essential performance of an aesthetic systems	This standard has been replaced by ISO 80601-2-13: 2022 – Medical electrical equipment: Parts 2–13: Particular requirements for basic safety and essential performance of an anaesthetic workstation. This standard is applicable to the basic safety and essential performance of an anaesthetic workstation for administering inhalation anesthesia. The standard specifies the particular requirements for a complete anaesthetic workstation and related components.
IEC 60601-2-16 Medical electrical equipment – Parts 2–16	Particular requirements for basic safety and essential performance of haemodialysis, hemodiafiltration and hemofiltration equipment.	This standard specifies the minimum safety requirements for hemodialysis equipment. These devices are intended for use by either medical staff or for use by the patient or other trained personnel under the supervision of medical expertise. This standard includes all electromedical equipment that is intended to deliver a haemodialysis, haemodiafiltration and haemodiafiltration treatment to a patient suffering from kidney failure.

IEC 60601-2-17 Medical electrical equipment – Parts 2–17	Particular requirements for the basic safety and essential performance of automatically controlled brachytherapy after loading equipment.	This standard applies to the basic safety and essential performance of automatically controlled Brachytherapy After loading Medical Equipment. The standard does not specify requirements for sealed radioactive sources.
IEC 60601-2-18 Medical electrical equipment – Parts 2–18	Particular requirements for the basic safety and essential performance of endoscopic equipment.	This standard specifies the particular basic safety and essential performance requirements for endoscopic equipment. The minimum safety requirements specified in this standard are considered for safety when it comes to endoscopic operation.
IEC 60601-2-19 Medical electrical equipment – Parts 2–19	Particular requirements for the basic safety and essential performance of infant incubators.	This standard specifies the safety and essential performance requirements for endoscopic equipment. It includes intense light source equipment which is part of the endoscopic equipment and its supply unit. This standard aids the following individuals or groups: • Manufacturers of endoscopic equipment. • Operators and technical staff. • Medical test laboratories. • Regulatory authorities.
IEC 60601-2-20 Medical electrical equipment – Parts 2–20	Particular requirements for the basic safety and essential performance of infant transport incubators.	This standard applies to the basic safety and essential performance of infant transport incubators. The standard does not apply to infant transport incubators, infant radiant warmers, devices supplying heat via blankets, pads, mattresses used in medical use and infant phototherapy equipment.

IEC 60601-2-21 Medical electrical equipment – Parts 2–21	Particular requirements for the basic safety and essential performance of infant radiant warmers.	This standard applies to the basic safety and essential performance of infant radiant warmers, also known as ME equipment. The standard specifies the safety requirements for infant radiant warmers but alternate methods of compliance with a specific clause through demonstrating safety equivalence. It should also be noted that the equipment will not be judged as non-compliant if the manufacturer can demonstrate that the risk presented by the hazard has been found to be of an acceptable level when weighed against the benefit of treatment using the device. This should be demonstrated in the manufacturers risk file. This particular standard does not apply to the following: • devices supplying heat using blankets, pads, mattresses in medical use • infant incubators • infant transport incubators • infant phototherapy equipment
IEC 60601-2-22 Medical electrical equipment – Parts 2–22	Particular requirements for basic safety and essential performance of surgical, cosmetic, therapeutic and diagnostic laser equipment.	This standard applies to the safety and essential performance of laser equipment for surgical, therapeutic, medical diagnostic, cosmetic, or veterinary applications which are intended for use on humans or animals. These devices are classified as a class 3B or class 4 laser product. This standard can also be applied to surgical, cosmetic, therapeutic and diagnostic laser equipment used for compensation or alleviation of disease, injury or disability.

IEC 60601-2-23 Medical electrical equipment – Parts 2–23	Particular requirements for the basic safety and essential performance of transcutaneous partial pressure monitoring equipment.	This standard is applicable to the basic safety and performance of transcutaneous partial pressure monitoring equipment, irrespective of whether this equipment is stand alone or part of a system. It applies to transcutaneous monitors used with adults, children and neonates. It also includes the use of these devices in fetal monitoring during birth. The standard does not apply to hemoglobin saturation oximeters or to devices applied to surfaces of the body other than the skin.
IEC 60601-2-24 Medical electrical equipment – Parts 2–24:	Particular requirements for the basic safety and essential performance of infusion pumps and controllers.	This standard is applicable to the basic safety and essential performance of infusion pumps and volumetric infusion controllers. It is applicable to administration sets as their characteristics tend to influence the basic safety and/or essential performance of infusion pumps and infusion controllers. The standard specifies the requirements for: • enteral nutrition pumps • infusion pumps • infusion pumps for ambulatory use • syringe or container pumps • volumetric infusion controllers • volumetric infusion pumps This standard does not specify the requirements for other aspects of administration tests. This includes testing. The standard does not apply to the following: • devices that are specifically intended for diagnostic or similar use

		• devices for extracorporeal circulation of blood • implantable devices • equipment specifically intended for diagnostic use within urodynamics • equipment specifically intended for diagnostic within male impotence testing • devices covered by ISO 28620
IEC 60601-2-25 Medical electrical equipment – Parts 2–25	Particular requirements for the basic safety and essential performance of electrocardiographs	This standard applies to the basic safety and essential performance of electrocardiographs as part of a medical electrical system. This is used for the production of electrocardiogram reports for diagnostic purposes. This equipment is intended for use under extreme or uncontrolled environments and conditions that are outside the hospital such as in ambulances and air transport. This standard is not applicable to: • vectorcardiography loops • ambulatory electrocardiographic medical electrical equipment which is not intended for obtaining electrocardiogram reports for diagnostic purposes • cardiac monitors which are not intended for obtaining electrocardiographic reports for diagnostic purposes
IEC 60601-2-26 Medical electrical equipment – Parts 2–26	Particular requirements for the basic safety and essential performance of electroencephalographs	This standard applies to the basic safety and essential performance of electroencephalographs which are used in a clinical environment such as the hospital, physicians' office and so on. The standard does not cover the requirements for other equipment used in electroencephalography.

IEC 60601-2-27 Medical electrical equipment – Parts 2–27	Particular requirements for the basic safety and essential performance of electrocardiographic monitoring equipment	This standard applies to electrocardiographic monitoring equipment used in a hospital environment as well as when used outside the hospital environment, such as in ambulances and air transport. The standard also covers electrocardiographic monitoring equipment intended for use under extreme or uncontrolled environmental conditions outside the hospital environment, such as in ambulances and air transport. This standard is not applicable to electrocardiographic monitors for home use. In addition, manufacturers should consider using relevant clauses of this standard as appropriate for their intended use.
IEC 60601-2-28 Medical electrical equipment – Parts 2–28	Particular requirements for the basic safety and essential performance of X-ray tube assemblies for medical diagnosis.	This standard is applicable to the basic safety and essential performance requirements for X-ray tube assemblies for medical diagnosis.
IEC 60601-2-29 Medical electrical equipment – Parts 2–29:	Particular requirements for the basic safety and essential performance of radiotherapy simulators.	This standard is applicable to the basic safety and essential performance of radiotherapy simulators which: • uses diagnostic x-ray equipment to simulate a therapeutic radiation beam physically such that the treatment volume subjected to irradiation during radiotherapy can be localized • the position and size of the radiotherapy radiation field can be confirmed The standard also establishes the requirements to ensure the ionizing radiation safety and enhanced mechanical and electrical safety of radiotherapy simulators.

		In addition, the standard identifies geometrical parameters that are critical for the accurate simulation of a radiotherapy treatment. It should also be noted that the standard does not intended to define optimum performance requirements. The standard does place limits on degradation of medical electrical equipment performance.
IEC 60601-2-31 Medical electrical equipment – Parts 2–31:	Particular requirements for the basic safety and essential performance of external cardiac pacemakers with internal power source	This standard applies to the basic safety and essential performance of external pacemakers which are powered by an internal electrical power source. The standard applies to patient cables but does not, however apply to patient leads. It should also be noted that this standard does not apply to active implantable medical devices – these are covered in another standard: ISO 14708-1. The standard also does not apply to external pacemakers which can be connected directly or indirectly to a supply main. In addition, the document does not apply to transthoracic and esophageal pacing ME equipment.
IEC 60601-2-33 Medical electrical equipment – Parts 2–33:	Particular requirements for the basic safety and essential performance of magnetic resonance equipment for medical diagnosis.	This standard applies to the basic safety and essential performance of magnetic resonance equipment and magnetic resonance systems for medical diagnosis. This standard does not cover the application of medical resonance equipment beyond the intended use.

IEC 60601-2-34 Medical electrical equipment – Parts 2–34:	Particular requirements for the basic safety and essential performance of invasive blood pressure monitoring equipment.	This standard applies to the basic safety and essential performance of invasive blood pressure monitoring equipment. The standard does not apply to: • catheter tubing • catheter needles • Luer locks • taps • tap tables • noninvasive blood pressure monitoring equipment
IEC 60601-2-36 Medical electrical equipment – Parts 2–36:	Particular requirements for the basic safety and essential performance of equipment for extracorporeally induced lithotripsy.	This standard applies to the basic safety and essential performance of equipment for extracorporeally induced lithotripsy. This includes equipment for other medical applications of therapeutic extracorporeally induced focused pressure pulses. This standard is limited to those components which are directly involved in the lithotripsy treatment. This includes: • the generator of the pressure pulse • patient support device • their interactions with imaging and monitoring devices The standard does not apply to other devices, such as patient treatment planning computers, X-ray and ultrasonic devices because they are treated in other applicable standards. In addition, this particular standard does not apply to ultrasound physiotherapy equipment intended to be used for physiotherapy and to ultrasound equipment intended to be used for high intensity therapeutic ultrasound and other therapy equipment.

IEC 60601-2-37 Medical electrical equipment – Parts 2–37:	Particular requirements for the basic safety and essential performance of ultrasonic medical diagnostic and monitoring equipment	This standard applies to the basic safety and essential performance of ultrasonic diagnostic equipment.
IEC 60601-2-39 Medical electrical equipment – Parts 2–39:	Particular requirements for basic safety and essential performance of peritoneal dialysis equipment.	This standard applies to the basic safety and essential performance of peritoneal dialysis equipment intended for use either by medical staff or under supervision of medical experts. This includes peritoneal dialysis equipment which may be operated by the patient regardless of whether the peritoneal dialysis equipment is used in a hospital or domestic environment.
IEC 60601-2-40 Medical electrical equipment – Parts 2–40:	Particular requirements for the basic safety and essential performance of electromyographs and evoked response equipment.	This standard applies to the basic safety and essential performance of electromyographs and evoked response equipment intended for transcutaneous electrical nerve stimulators.
IEC 60601-2-41 Medical electrical equipment – Parts 2–41:	Particular requirements for the basic safety and essential performance of surgical luminaires and luminaires for diagnosis.	This standard applies to the basic safety and essential performance of luminaires and luminaires for diagnosis. The standard does not apply to: • headlights • endoscopes • laparoscopes and their light sources • luminaires used in dentistry • luminaires for general purpose • luminaires detected for therapeutic purposes

		• special purpose lights with different conditions of use (such as light sources intended for decontamination of air/surfaces, UV lights for dermatological diagnosis, slit lamps for ophthalmology, lights for surgical microscopes and lights for surgical navigation systems, lights connected to surgical instruments and luminaires for emergency lighting.
IEC 60601-2-43 Medical electrical equipment – Parts 2–43:	Particular requirements for the basic safety and essential performance of X-ray equipment for interventional procedures.	This standard applies to the basic safety and essential performance of both fixed and mobile x-ray equipment declared by the manufacturer to be suitable for radioscopically guided interventional procedures. The standard scope excludes: • equipment for radiotherapy • equipment for computed tomography • accessories intended to be introduced into the patient • mammographic x-ray equipment • dental x-ray equipment
IEC 60601-2-44 Medical electrical equipment – Parts 2–44:	Particular requirements for the basic safety and essential performance of X-ray equipment for computed tomography.	This standard applies to the basic safety and essential performance of x-ray equipment for computed tomography. The scope of the standard is limited to CT scanners which are intended to used for both head and body, and which is characterized by an enclosure of the x-ray source(s) and imaging detector(s) in a common protective cover. The standard also includes the safety requirements for the x-ray generators and CT scanners.
IEC 60601-2-45 Medical electrical equipment – Parts 2–45:	Particular requirements for basic safety and essential performance of mammographic X-ray equipment and mammographic stereotactic devices.	This standard applies to the basic safety and essential performance of mammographic stereotactic devices. The standard addresses the system level of mammographic X-ray equipment, which consists of a combination of an X-ray generator, associated equipment and accessories.

IEC 60601-2-46 Medical electrical equipment – Parts 2–46	Particular requirements for the basic safety and essential performance of operating tables.	This standard specifies the basic safety and essential performance requirements for operating tables. This standard does not apply to dental patient chairs, examination chairs, couches, patient supporting systems of diagnostic, interventional and therapeutic equipment, operating table heating blankets, patient transfer equipment, delivery table equipment, delivery tables and delivery beds, medical beds and field tables.
IEC 60601-2-47 Medical electrical equipment – Parts 2–47	Particular requirements for the basic safety and essential performance of ambulatory electrocardiographic systems.	This standard provides the particular requirements for the basic safety and essential performance of ambulatory electrocardiographic systems. The standard covers the following: • systems that provide continuous recording and continuous analysis of the ECG allowing full re-analysis. • systems that provide continuous analysis and only partial or limited recording not allowing a full re-analysis of the ECG. In addition, the safety aspects of this standard apply to all types of systems falling into the above-mentioned categories.
IEC 60601-2-49 Medical electrical equipment – Parts 2–49:	Particular requirements for the basic safety and essential performance of multifunction patient monitoring equipment.	This standard applies to the basic safety and essential performance of multifunction patient monitoring equipment. In addition, it also applies to medical electric equipment which is used in a hospital environment as well as outside the hospital environment such as in ambulances and air transport.

IEC 60601-2-50 Medical electrical equipment – Parts 2–50:	Particular requirements for the basic safety and essential performance of infant phototherapy equipment.	This standard applies to the basic safety and essential performance of infant phototherapy equipment. The standard does not cover any hazards inherent in the physiological function of the equipment or related systems. It should also be noted that alternate methods of compliance will be regarded as being non-compliant if the manufacturer can demonstrate in his/her risk management profile that the risk presented by the hazard has been found to be acceptable when weighed against the benefit of treatment. The standard does not apply to: • devices supplying heat via blankets • pads and/or mattresses in medical use • infant incubators • infant transport incubators • infant radiant warmers
IEC 60601-2-52 Medical electrical equipment – Parts 2–52:	Particular requirements for the basic safety and essential performance of medical beds.	This standard applies to the basic safety and essential performance of medical beds intended for adults.
IEC 60601-2-54 Medical electrical equipment – Parts 2–54:	Particular requirements for the basic safety and essential performance of X-ray equipment for radiography and radioscopy.	This standard applies to the basic safety and essential performance of medical electrical equipment and medical electrical systems intended to be used for projection radiography and radioscopy. The standard aims to provide a practical degree of safety when the device is in operation.
IEC 60601-2-57 Medical electrical equipment – Parts 2–57	Particular requirements for the basic safety and essential performance of non-laser light source equipment intended for therapeutic, diagnostic, monitoring and cosmetic/aesthetic use.	This standard applies to the basic safety and essential performance of equipment which incorporates more that one source of optical radiation in the wavelength of 200 nm to 3000 nm (with the exception of laser radiation) and which is intended to create photobiological effects in humans for therapeutic, diagnostic, monitoring, and cosmetic or aesthetic applications.

IEC 60601-2-62 Medical electrical equipment – Parts 2–62:	Particular requirements for the basic safety and essential performance of high intensity therapeutic ultrasound (HITU) equipment.	This standard applies to the basic safety and essential performance of high intensity therapeutic ultrasound equipment. The standard does not include hazards inherent in the intended physiological function of the equipment or systems.
IEC 60601-2-63 Medical electrical equipment – Parts 2–63:	Particular requirements for the basic safety and essential performance of dental extra-oral X-ray equipment.	This standard applies to the basic safety and essential performance of dental extra-oral X-ray equipment.
IEC 60601-2-64 Medical electrical equipment – Parts 2–64:	Particular requirements for the basic safety and essential performance of light ion beam medical electrical equipment.	This standard applies to the basic safety and essential performance of light ion bean medical electrical equipment. This standard applies specifically to the manufacture and installation aspects of ME equipment intended for radiotherapy in human medical practice. This includes those in which the selection and display of operating parameters can be controlled automatically by programmable electronic subsystems, that, in normal use, deliver a radiation beam of light ions having energy per nucleon in the range 10 MeV/n to 500 MeV/n.
IEC 60601-2-65 Medical electrical equipment – Parts 2–65	Particular requirements for the basic safety and essential performance of dental intra-oral X-ray equipment.	This standard applies to the basic safety and essential performance of dental intra-oral X-ray equipment and its main components. The standard is restricted to X-ray equipment where the X-ray tube assembly contains the high-voltage transformer assembly. The standard is not applicable to dental extra-oral X-ray equipment.

IEC 60601-2-66 Medical electrical equipment – Parts 2–66:	Particular requirements for the basic safety and essential performance of hearing instruments and hearing instrument systems.	This standard applies to the basic safety and essential performance of hearing instruments/hearing aids and hearing instrument systems/hearing aid systems.
IEC 60601-2-68 Electrical medical equipment – Parts 2–68:	Particular requirements for the basic safety and essential performance of X-ray-based image-guided radiotherapy equipment for use with electron accelerators, light ion beam therapy equipment and radionuclide beam therapy equipment.	This standard applies to the basic safety and essential performance of x-ray-based image-guided radiotherapy equipment for use with electron accelerators, light ion beam therapy equipment and radionuclide beam therapy equipment. The standard covers the safety aspects of kilo voltage and mega voltage x-ray imaging devices as well as aspects of communication between EBE and x-ray imaging devices.
IEC 60601-2-75 Medical electrical equipment – Parts 2–75	Particular requirements for the basic safety and essential performance of photodynamic therapy and photodynamic diagnosis equipment.	This standard applies to the basic safety and essential performance of photodynamic therapy and photodynamic diagnosis equipment. The standard applies to photodynamic therapy and photodynamic diagnosis equipment which is used for compensation or alleviation of disease, injury or disability. The standard does not cover hazards inherent in the intended physiological function of the equipment. In addition, this standard does not apply to: • light therapy equipment intended for use in photothermal ablation, coagulation, and hyperthermia; • low-level laser therapy equipment not intended for use with a photosensitizer. • illumination equipment intended for use in observation, monitoring, and diagnosis, not intended for use with a photosensitizer.

IEC 60601-2-76:2018 Medical electrical equipment – Parts 2–76:	Particular requirements for the basic safety and essential performance of low energy ionized gas hemostasis equipment.	This standard applies to: • the basic safety and essential performance of low energy ionized gas hemostasis equipment. • The standard does not cover hazards inherent in the intended physiological function of equipment or system.
IEC 60601-2-83:2019 Medical electrical equipment – Parts 2–83:	Particular requirements for the basic safety and essential performance of home light therapy equipment.	This standard is applicable to the basic safety and essential performance of home light therapy equipment which is intended for use in the home healthcare environment. The standard covers all light sources except laser.
IEC 60601-2-84 Medical electrical equipment – Parts 2–84:	Particular requirements for the basic safety and essential performance of emergency and transport ventilators.	This standard applies to the basic safety and essential performance of emergency and transport ventilators as well as its accessories. This is particularly in cases where the equipment is: • intended for patients who need differing levels of support from artificial ventilation. • intended to be operated by a healthcare professional. • intended for use in the EMS environment. • intended for invasive or non-invasive ventilation.

The IEC 60601 series of standards is aimed at ensuring the safety of patients using medical electrical equipment. Compliance with these standards helps to minimize the risk of electrical hazards, such as electric shocks or burns, ensuring that the medical device is designed and manufactured to prioritize patient safety. The series of standards also provides a consistent framework for safety requirements across different types of medical electrical equipment.

4.15. UNDERSTANDING IEC 62366: STANDARDS FOR USABILITY ENGINEERING

Usability engineering, also known as human factors engineering, focuses on designing a user interface that allows people to quickly understand how to interact with a product in the most efficient and error-free manner. In the medical device industry, usability engineering has become important to ensure that the design of the medical device considers the user-device interaction to ensure both patient and user safety. The ultimate goal of usability engineering is to identify and mitigate any use-related hazards and risks, and to create a user interface that encourages error-free use of the device.

IEC 62366 is the international standard that covers the application of usability engineering to medical devices. This standard helps medical device manufacturers consider human factors by guiding a standardized process for conducting usability analyzes of their medical devices.

IEC 62366 is broken into two sub-standards:

- **IEC 62366-1– Part 1:** Application of usability engineering to medical devices
- **IEC 62366-2 – Part 2:** Guidance on the application of usability engineering to medical devices

The standard outlines a series of steps to ensure that the usability interface of a medical device has been thoroughly tested. Critical steps include:

- Defining the intended users, use environments, and user interface;
- Identification of use-related hazards;
- Identification and categorization of critical tasks;
- Development and implementation of risk mitigation/control measures;
- Validation of user safety and effectiveness;
- Creation of feedback loops if validation shows that use-related risks are unacceptable, or new use-related risks have been introduced;
- Requirements for documenting the usability engineering process.

The standard also provides details on conducting usability engineering validation through two main forms: formative evaluations and summative evaluations.

- **Formative Evaluations:** These are conducted during the early stages of design and development and are used to address safety concerns that may appear during preliminary analyzes.
- **Summative Evaluations:** These are conducted during the design validation stage and form a more detailed evaluation. The goal of summative evaluations is to show that the device is indeed safe when used. This stage often involves testing the final design with the actual intended users under realistic conditions.

Key considerations during usability engineering include:

- An analysis of the interaction between the intended user and the device.
- The information users perceive from the device (including labeling, packaging, and instructions for use).
- How users interpret information when making decisions about device operation.
- How the device is used.
- How the device receives inputs from the user.
- How the device responds to the user and provides feedback.
- Any design issues, including software or hardware.

4.16. UNDERSTANDING ISO 10993: STANDARDS FOR BIOCOMPATIBILITY

The ISO 10993 series of standards focuses on the biological evaluation of medical devices. These standards provide guidelines for assessing the potential risks associated with the use of medical devices based on their type and duration of contact.

Types of contact include:

- **Non-Contacting Medical Devices**: Devices that do not make direct or indirect contact with patients.
- **Surface-Contacting Devices**: Devices that touch the skin, intact mucous membranes, and breached or compromised surfaces.
- **Externally Communicating Devices**: Devices that are partially or completely external and come into contact with bodily fluids. These devices are usually intended to deliver or draw fluids to or from the body and are attached to an external delivery or withdrawal system.
- **Implantable Devices**: Devices embedded within human tissue.

Categorizations for duration of contact include:

- **Limited Exposure**: Devices with a cumulative contact time of up to 24 hours.
- **Prolonged Exposure**: Devices with a cumulative contact time exceeding 24 hours but not more than 30 days.
- **Long-Term Exposure**: Devices with a cumulative contact time exceeding 30 days.

Each standard covers a specific aspect of biocompatibility, such as cytotoxicity, sensitization, or irritation. Below is a list of all the relevant standards for the biocompatibility of medical devices:

ISO 10993-1	Evaluation and testing within a risk management process	(Latest 2020) This standard serves as an introduction to the ISO 10993 series, providing an overview of the biological evaluation of medical devices. It sets the foundation for subsequent parts, outlining general principles and considerations.
ISO 10993-2	Animal welfare requirements:	(Latest 2022) This standard guides the ethical treatment of animals in biocompatibility testing for medical devices. It emphasizes minimizing discomfort and ensuring humane practices during testing procedures.
ISO 10993-3	Tests for genotoxicity, carcinogenicity, and reproductive toxicity	(Latest 2014) Addressing potential long-term risks, this standard outline testing methods to evaluate genotoxic, carcinogenic, and reproductive effects associated with the use of medical devices.
ISO 10993-4	Selection of tests for interactions with blood:	(Latest 2017) Specifically tailored for devices in contact with blood, this standard provides criteria for selecting tests to assess the interactions between medical devices and blood components.
ISO 10993-5	Tests for in vitro cytotoxicity	(Latest 2009) Focused on cellular response, this standard details procedures for in vitro cytotoxicity testing, ensuring that medical devices do not induce harmful effects on cells.
ISO 10993-6	Tests for local effects after implantation	(Latest 2016) Addressing local effects post-implantation, this standard guides the evaluation of tissue responses to medical devices, emphasizing the importance of assessing biocompatibility at the implantation site.

ISO 10993-7	Ethylene oxide sterilization residuals	(Latest 2008) Centered on sterilization methods, this standard focuses on ethylene oxide residuals, ensuring that medical devices treated with this method meet safety standards regarding residual levels.
ISO 10993-9	Framework for identification and quantification of potential degradation products:	(Latest 2019) Addressing potential degradation, this standard establishes a framework for identifying and quantifying degradation products, ensuring the safety and efficacy of medical devices over time.
ISO 10993-10	Tests for skin sensitization	(Latest 2021) Focusing on skin irritation and sensitization, this standard outline testing methods to assess the potential for medical devices to cause skin-related adverse effects.
ISO 10993-11	Tests for systemic toxicity	(Latest 2018) This standard delves into systemic toxicity testing, ensuring a comprehensive evaluation of medical devices' impact on the entire organism. It addresses potential adverse effects beyond local responses, contributing to overall safety assessments.
ISO 10993-12	Sample preparation and reference materials	(Latest 2021) Focused on accuracy, this standard provides guidelines for sample preparation and the use of reference materials in biocompatibility testing. It emphasizes consistency and reliability in test procedures.
ISO 10993-13	Identification and quantification of degradation products from polymeric medical devices	(Latest 2019) Tailored for polymeric medical devices, this standard establishes methods for identifying and quantifying degradation products, crucial for ensuring the longevity and safety of such devices.

ISO 10993-14	Identification and quantification of degradation products from ceramics	(Latest 2009) This standard provides guidance on specific tests for the identification and quantification of the degradation products associated with the use of ceramics in medical devices. It ensures the long-term stability and biological safety of ceramic medical devices.
ISO 10993-15	Identification and quantification of degradation products from metals and alloys	(Latest 2023) This standard is focused on design of tests for identifying and quantifying degradation products from final metallic medical devices or corresponding material samples finished as ready for clinical use. The standard ensures that the degradation products do not pose any unacceptable risks to patients.
ISO 10993-16	Toxicokinetic study design for degradation products and leachables	(Latest 2017) Addressing the toxicokinetic aspects, this standard guides the design of studies focused on degradation products and leachables. It ensures a comprehensive understanding of how these substances interact with the body over time.
ISO 10993-17	Toxicological risk assessment of medical device constituents	(Latest 2023) This standard focuses on the toxicological risk assessment of medical device constituents. It ensures that the potential risks associated with the exposure of materials and processes used in manufacturing of the device are thoroughly evaluated.
ISO 10993-18	Chemical characterization of medical device materials within a risk management process.	(Latest 2023) Centered on material characterization, this standard provides guidelines for the chemical analysis of materials used in medical devices. It ensures a thorough understanding of the composition and potential risks associated with device materials.

ISO 10993-19	Physico-chemical, morphological, and topographical characterization of materials	(Latest 2020) Expanding on material characterization, this standard includes physico-chemical, morphological, and topographical aspects. It contributes to a comprehensive assessment of how these characteristics influence biocompatibility.
ISO/TS 10993-20	Principles and methods for immunotoxicology testing of medical devices.	(Latest 2006) Focused on immunotoxicology, this standard outlines principles and methods for assessing the immunological impact of medical devices. It ensures a thorough evaluation of potential immune system responses to device exposure.
ISO/TR 10993-22	Guidance on nanomaterials	(Latest 2017) This standard provides a general framework for assessing the safety of medical devices made of nanomaterials. It ensures a thorough toxicological evaluation of nanomaterials and their biological safety in medical devices.
ISO 10993-23	Tests for irritation	(Latest 2021) This standard is focused on the tests designed to predict and classify the irritation potential of medical devices, materials, or their extracts. It ensures a thorough evaluation of any potential irritation reactions on biological tissues, thereby contributing to the overall safety of medical devices.

4.17. UNDERSTANDING ISO 14971: MEDICAL DEVICES – APPLICATION OF RISK MANAGEMENT TO MEDICAL DEVICES

ISO 14971 is the international standard that defines how to apply risk management to medical devices. The first edition of ISO 14971 was published in 2000, and the current iteration, the third edition, was released in 2019. As the risk management standard has evolved, it has increasingly focused on integrating risk management into the quality management system to fully embed the risk process into the product life cycle. The most recent edition of ISO 14971 has added more requirements for post-market risk management, requiring manufacturers to analyze post-market data and better identify potential trends.

The purpose of this standard is to aid in the compliance of medical devices through the application of risk management guidelines while reducing uncertainties about product safety. The standard helps provide methods for all stakeholders to eliminate risks while helping to develop a device that performs effectively and is safe. ISO 14971 includes the following overall requirements:

- How to implement a risk management system;
- Risk management system review by management;
- Risk acceptability criteria and related policies;
- Definition of qualified personnel;
- Provision of adequate resources.

4.18. UNDERSTANDING THE DIFFERENCES BETWEEN THE 2007, 2012, AND 2019 VERSIONS OF THE STANDARD

Before delving deeper into ISO 14971, it's important to look at the changes made between the most recently published version (2019) and the previous versions.

The 2019 version of ISO 14971 was released in December 2019, replacing the two prior versions released in 2012 and 2007:

- ISO 14971:2007 (first international version);
- ISO 14971:2012.

The EN version of ISO 14971 for medical devices, similar to the regular version of ISO 14971, is only applicable in Europe and was published by the European Committee for Standardization with respect to the relevant regulations at the time, namely: European Directives 93/42/EEC, 98/79/EC, and 90/385/EEC.

ISO 14971:2019 is the third version of the standard with three major changes as follows:

1. **New definitions:** These for certain terminology have been introduced:
   i. **Benefit:** Described as a positive impact or outcome that is desirable while using a medical device.
   ii. **Harm:** Previously referred to as "physical injury." The word

"harm" was mentioned in the previous definition but has now been deleted, while injury or damage to public health, property, and the environment is mentioned.

iii. **Reasonably Foreseeable Misuse:** This term outlines the use of a product in a way that is not intended by the manufacturer. Misuse can either be unintentional or intentional, which may also be referred to as abnormal use.

iv. **State of the Art:** Risk policies are based on the state of the art, and ISO 14971:2019 now provides a definition of this terminology.

2. **Criteria for Overall Risk Acceptability:** The 2019 version suggests that in comparison to overall residual risks, manufacturers may use different evaluation criteria for residual risk of individual risks.

3. **More Information in the Guidance to the Technical Report (ISO 24971):** The new version of ISO 14971:2019 includes three annexes with extensive information and explanations. This has been further elaborated on in ISO 24971 (Tables 4.1 and 4.2).

Table 4.1. Comparison between ISO 14971:2007 and ISO 14971:2019

ISO 14971:2007	ISO 14971:2019
Introduction	Introduction
1. Scope	1. Scope
(new clause)	2. Normative references
2. Terms and definitions	3. Terms and definitions
2.1. Accompanying document	3.1. Accompanying document
(new definition)	3.2. Benefit
2.2. Harm	3.3. Harm
2.3. Hazard	3.4. Hazard
2.4. Hazardous situation	3.5 Hazardous situation
2.5. Intended use Intended purpose	3.6. Intended use Intended purpose
2.6. In vitro diagnostic medical device IVD medical device	3.7. In vitro diagnostic medical device IVD medical device
2.7. Life cycle	3.8. Life cycle
2.8. Manufacturer	3.9. Manufacturer
2.9. Medical device	3.10. Medical device

2.10. Objective evidence	3.11. Objective evidence
2.11. Post-production	3.12. Post production
2.12. Procedure	3.13. Procedure
2.13. Process	3.14. Process
(new definition)	3.15. Reasonably foreseeable misuse
2.14. Record	3.16. Record
2.15. Residual risk	3.17. Residual risk
2.16. Risk	3.18. Risk
2.17. Risk analysis	3.19. Risk analysis
2.18. Risk assessment	3.20. Risk assessment
2.19. Risk control	3.21. Risk control
2.20. Risk estimation	3.22. Risk estimation
2.21. Risk evaluation	3.23. Risk evaluation
2.22. Risk management	3.24. Risk management
2.23. Risk management file	3.25. Risk management file
2.24. Safety	3.26. Safety
2.25. Severity	3.27. Severity
(new definition)	3.28. State of the art
2.26. Top management	3.29. Top management
2.27. Use error	3.30. Use error
2.28. Verification	3.31. Verification
3. General requirements for risk management	4. General requirements for risk management system
3.1. Risk management process	4.1. Risk management process
3.2. Management responsibilities	4.2. Management responsibilities
3.3. Qualification of personnel	4.3. Competence of personnel
3.4. Risk management plan	4.4. Risk management plan
3.5. Risk management file	4.5. Risk management file
4. Risk analysis	5. Risk analysis
4.1. Risk analysis process	5.1. Risk analysis process
4.2. Intended use and identification of characteristics related to the safety of the medical device	5.2. Intended use and reasonably foreseeable misuse 5.3. Identification of characteristics related to safety
4.3. Identification of hazards	5.4. Identification of hazards and hazardous situations

4.4. Estimation of the risk(s) for each hazardous situation	5.5. Risk estimation
5. Risk evaluation	6. Risk evaluation
6. Risk control	7. Risk control
6.1. Risk reduction	(subclause deleted)
6.2. Risk control option analysis	7.1. Risk control option analysis
6.3. Implementation of risk control measure(s)	7.2. Implementation of risk control measures
6.4. Residual risk evaluation	7.3. Residual risk evaluation
6.5. Risk/benefit analysis	7.4. Benefit-risk analysis
6.6. Risks arising from risk control measures	7.5. Risks arising from risk control measures
6.7. Completeness of risk control	7.6. Completeness of risk control
7. Evaluation of overall residual risk acceptability	8. Evaluation of overall residual risk
8. Risk management report	9. Risk management review
9. Production and post-production information	10. Production and post-production activities 10.1. General 10.2. Information collection 10.3. Information review 10.4. Actions
Annex A Rationale for requirements	Annex A Rationale for requirements
Annex B Overview of the risk management process for medical devices	Annex B Risk management process for medical devices
Annex C Questions that can be used to identify medical device characteristics that could impact on safety	Moved to ISO/TR 24971
Annex D Risk concepts applied to medical devices	
Annex E Examples of hazards, foreseeable sequences of events and hazardous situations	Annex C Fundamental risk concepts
Annex F Risk management plan	Moved to ISO/TR 24971
Annex G Information on risk management techniques	
Annex H Guidance on risk management techniques	

Annex I Guidance on risk analysis process for biological hazards	Annex deleted
Annex J Information for safety and information about residual risk	Moved to ISO/TR 24971
Bibliography	Bibliography

Table 4.2. *Comparison between ISO 14971:2012 and ISO 14971:2019*

ISO 14971:2007/EN 14971:2012	ISO 14971:2019
1. Scope	1. Scope
2. Terms and definitions	2. Normative references
3. Risk management requirements: • Process of risk management • Responsibilities • Personnel qualification • Plan for management of risk • File for risk management	3. Definitions and terms
4. Analysis of risk: • Process of risk analysis • Identification of safety characteristics and intended use of medical device • Hazards identification • Estimation of each hazardous situation for risk	4. Risk management requirements: • Process of risk management • Responsibilities • Personnel qualification • Plan for management of risk • File for risk management
5. Evaluation of risk	5. Analysis of risk: • Process of risk analysis • Reasonable foreseeable misuse and intended use • Identification of safety characteristic • Identification of hazardous situations or hazards • Estimation of risk

6. Control of risk • Reduction of risk • Analysis of risk control options • Implementation of measures for risk control • Evaluation of residual risk • Analysis of benefit risk • Arising risks from measures for risk control • Risk control completeness	7. Evaluation of risk
7. Overall residual risk acceptability evaluation	7. Control of risk • Analysis of risk control option • Implementation of measures for risk control • Evaluation of residual risk • Analysis of benefit-risk • Arising risks from measures for risk control • Risk control completeness
8. Report for risk management	8. Evaluation of the overall residual risk
9. Production – Post production activities	9. Risk management review
	9. Production – post-production activities: • Collection of information • Review of information • Actions

The standard also provides an overview of the risk management process which is summarized in Figure 4.1.

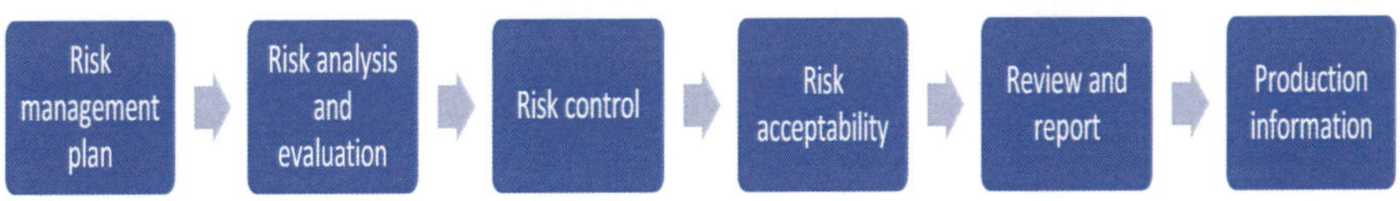

Figure 4.1. *Steps in the risk management process of medical devices.*

4.19. UNDERSTANDING THE ISO 14971 STANDARD IN DETAIL

The ISO 14971 standard defines the scope by indicating that it includes a process for risk management of medical devices and who the standard is intended for. It also specifies which phases of the medical device life cycle the standard applies to, as well as what it does not cover. The scope is defined as follows:

> 1. **Scope:** This document specifies terminology, principles, and a process for risk management of medical devices, including software as a medical device and in vitro diagnostic medical devices. The process described in this document assists manufacturers in identifying hazards associated with the medical device, estimating and evaluating the associated risks, controlling these risks, and monitoring the effectiveness of the controls.
>
> The requirements of this document apply to all phases of a medical device's life cycle. The process described applies to risks associated with a medical device, such as those related to biocompatibility, data and systems security, electricity, moving parts, radiation, and usability.
>
> The process can also be applied to products that are not necessarily medical devices in some jurisdictions and may be used by others involved in the medical device life cycle.
>
> This document does not apply to:
>
> - Decisions on the use of a medical device in the context of any particular clinical procedure; or
> - Business risk management.
>
> This document requires manufacturers to establish objective criteria for risk acceptability but does not specify acceptable risk levels.
>
> Risk management can be an integral part of a quality management system. However, this document does not require the manufacturer to have a quality management system in place.
>
> *Note: Guidance on the application of this document can be found in ISO/TR 24971.*

The standard then discusses the general requirements for risk management systems, including how the risk management process should be carried out. Section 4 of the standard states that the manufacturer should establish, implement, document, and maintain a risk management process. This process should assist in identifying hazards and related situations associated with the device. It should also estimate and evaluate associated risks, provide guidance on controlling these risks, and monitor the effectiveness of any implemented risk control measures.

This section also specifies that the process should apply throughout the life cycle of the device and include risk analysis, risk evaluation, risk control, and production and post-production activities.

The manufacturer is required to establish a risk management process as part of the design and development of a medical device. Risk analysis, risk evaluation, and risk control are essential parts of the risk management process. It is also important to understand that the risk management process continues post-production as well.

In addition, when a manufacturer employs a quality management system, this risk management process should be fully integrated into the quality management system.

4. General Requirements for Risk Management System

4.1. Risk Management Process

The manufacturer shall establish, implement, document, and maintain an ongoing process for:

- Identifying hazards and hazardous situations associated with a medical device;
- Estimating and evaluating the associated risks;
- Controlling these risks; and
- Monitoring the effectiveness of the risk control measures.

This process shall apply throughout the life cycle of the medical device and shall include the following elements:

- Risk analysis;
- Risk evaluation;
- Risk control; and
- Production and post-production activities.

Where a documented product realization process exists, it shall incorporate the appropriate parts of the risk management process.

Section 4 of the standard also touches on product realization processes. The manufacturer should have adequate processes in place that aid product realization. Guidance on product realization is given in Section 7 of the ISO 13485 standard. Product realization describes a range of processes involved in a product's life cycle, from conception to completion, including design, development, manufacturing, packaging, shipping, and maintenance. Product realization is mandatory as part of the Quality Management System (Figure 4.2).

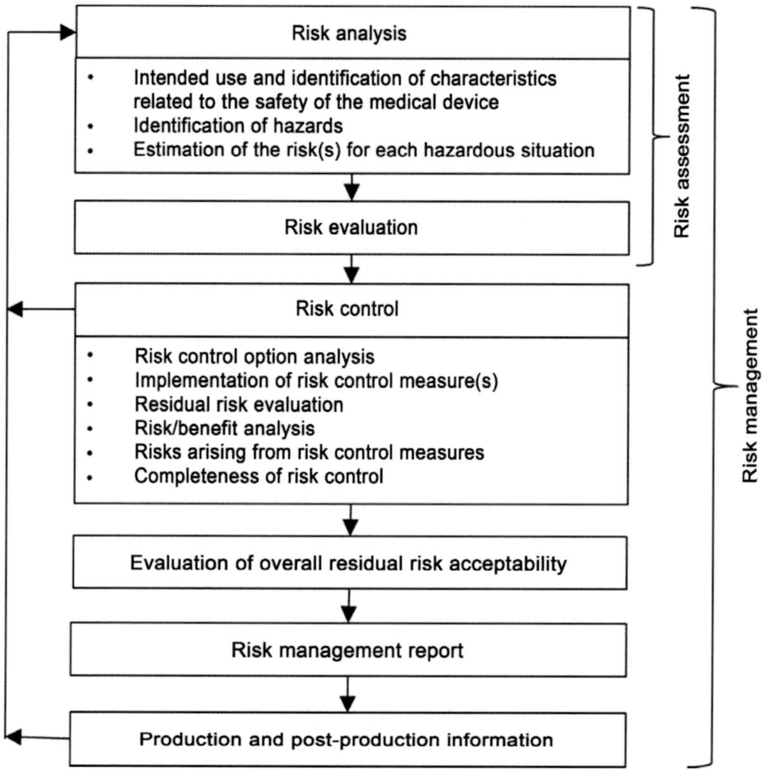

Figure 4.2. *Schematic representation of the risk management process.*

Section 4.2 of the ISO 14971:2019 standard focuses on the responsibilities of management and their commitment to the risk management process. The standard requires that this commitment is demonstrated through the provision of adequate resources and the assignment of competent personnel. In addition, the standard specifies that top management shall define and document a policy for risk management that includes relevant national regulations and international standards.

A risk policy template is provided at the end of this book.

The commitment of top management is crucial for an effective risk management process. Top management is responsible for providing overall guidance, resources, and competent personnel. Without adequate resources, risk management activities would be less effective, even if other requirements of the document are met. Furthermore, competent individuals trained in risk management techniques are essential. ISO 14971 does not define acceptable risk levels, so top management must establish a policy for determining acceptable risks. Periodic review of risk management activities is also necessary to ensure they are being carried out correctly, to identify weaknesses, implement improvements, and adapt to changes.

> 4.2. Management Responsibilities
>
> Top management shall provide evidence of its commitment to the risk management process by ensuring:
>
> - the provision of adequate resources; and
> - the assignment of competent personnel for risk management.
>
> Top management shall define and document a policy for establishing criteria for risk acceptability. The policy shall provide a framework that ensures criteria are based on applicable national or regional regulations, relevant International Standards, and available information, such as the generally acknowledged state of the art and known stakeholder concerns.
>
> Ø **Note 1:** The manufacturer's policy for establishing criteria for risk acceptability can define approaches to risk control, such as reducing risk as low as reasonably practicable (ALARP), as low as reasonably achievable (ALARA), or reducing risk as far as possible without adversely affecting the benefit-risk ratio.
>
> Top management shall review the suitability of the risk management process at planned intervals to ensure its continuing effectiveness and document any decisions and actions taken. If the manufacturer has a quality management system in place, this review may be part of the quality management system review.
>
> Ø **Note 2:** The results of reviewing production and post-production information can serve as input for the review of the risk management process.
>
> Ø **Note 3:** The documents described in this subclause can be incorporated into the documents produced by the manufacturer's quality management system and referenced in the risk management file.

Basic requirements for top management include:

- Provision of adequate resources;
- Assignment of competent personnel involved in the risk management process;

- Definition and documentation of a risk policy, including criteria for risk acceptability;
- Periodic review of the suitability of the risk management process;
- Documentation of decisions and actions taken regarding the risk management process.

Section 4.3 of the standard outlines the requirements for establishing personnel competence. Competence shall be based on education, training, skills, and/or experience. Personnel should also have knowledge of and experience with particular medical devices.

A competent risk management team inclusive of people with the knowledge and experience necessary to perform risk management tasks is imperative to the business. The risk management process requires people with knowledge and experience in the following:

- Who understand how the medical device is constructed;
- Who understand how the medical device works;
- Who understand how the medical device is produced;
- Who understand how the medical device is actually used;
- Who understand how to apply the risk management process.

Often, a multidisciplinary team from various functions or disciplines is needed. Each team member should contribute their specialist knowledge, and the balance and relationships between representatives should be considered. While records are required to provide evidence of competence, these do not need to be maintained in the risk management file due to confidentiality and data protection requirements.

4.3. Competence of Personnel

Persons performing risk management tasks shall be competent based on education, training, skills, and experience appropriate to the tasks assigned to them. Where appropriate, these persons shall have knowledge of and experience with the particular medical device (or similar devices) and its use, the technologies involved, or the risk management techniques employed. Appropriate records shall be maintained.

Ø **Note:** Risk management tasks can be performed by representatives from various functions, each contributing specialist knowledge.

Section 4.4 of the standard addresses the requirements for establishing and documenting a risk management plan. The plan shall be part of the risk management file.

A risk management plan is necessary because it provides an organized approach to risk management and serves as a roadmap for the process. It encourages objectivity and ensures that no essential elements are overlooked. The standard lists essential elements of the plan in section 4.4 (a-g), explained below:

- The scope of the planned risk management activities, identifying and describing the medical device and the life cycle phases for which each element of the plan is applicable.

The plan should cover both the medical device and the relevant phases of its life cycle. Defining the scope establishes the baseline for all risk management activities.

- Assignment of responsibilities and authorities. Responsibilities and authorities should be allocated to ensure no aspect of the risk management process is overlooked.
- Requirements for the review of risk management activities. Review of risk management activities is a general management responsibility.
- Criteria for risk acceptability, based on the manufacturer's policy for determining acceptable risk, including criteria for accepting risks when the probability of harm cannot be estimated. Criteria for risk acceptability should be established before risk analysis begins.
- A method to evaluate the overall residual risk and criteria for acceptability of the overall residual risk, based on the manufacturer's policy for determining acceptable risk.

After implementing all risk control measures, the overall impact of residual risks should be evaluated.

- Activities for verification of the implementation and effectiveness of risk control measures.

Verification ensures that essential resources are available when needed and that important parts of the verification process are not neglected.

- Activities related to the collection and review of relevant production and post-production information.

Methods for collecting and reviewing production and post-production information should be established to ensure relevant data is fed back into the risk management process. A record of changes should be maintained to facilitate audit and review.

4.4. Risk Management Plan

Risk management activities shall be planned. For the particular medical device, the manufacturer shall establish and document a risk management plan as part of the risk management file. The plan shall include at least the following:

- The scope of the planned risk management activities, identifying and describing the medical device and the life cycle phases for which each element of the plan is applicable;
- Assignment of responsibilities and authorities;
- Requirements for the review of risk management activities;
- Criteria for risk acceptability, based on the manufacturer's policy for determining acceptable risk, including criteria for accepting risks when the probability of harm cannot be estimated;

Note 1: The criteria for risk acceptability are essential for the ultimate effectiveness of the risk management process. For each risk management plan the manufacturer needs to establish risk acceptability criteria that are appropriate for the particular medical device.

- A method to evaluate the overall residual risk and criteria for acceptability of the overall residual risk, based on the manufacturer's policy for determining acceptable risk;

Note 2: The method to evaluate the overall residual risk can include gathering and reviewing data and literature for the medical device being considered and similar medical devices on the market and can involve judgment by a cross-functional team of experts with application knowledge and clinical expertise.

- Activities for verification of the implementation and effectiveness of risk control measures; and
- Activities related to the collection and review of relevant production and post-production information.

Note 3: See ISO/TR 24971 for guidance on developing a risk management plan and on establishing criteria for risk acceptability.

Note 4: Not all parts of the plan need to be created at the same time. The plan or parts of it can be developed over time.

If the plan changes during the life cycle of the medical device, a record of the changes shall be maintained in the risk management file.

An overview of what a risk management plan shall contain is provided below:

- Scope of the planned risk management activities.
- Identification and description of the medical device.
- Identification and description of the life cycle phases for which each element of the plan is applicable.
- Assignment of responsibilities and authorities.
- Requirements for reviewing risk management activities.
- Criteria for risk acceptability (a separate risk plan can be established for each medical device or device family. For each plan, the manufacturer must establish appropriate risk acceptability criteria).
- Method to evaluate the overall residual risk.
- Criteria for the acceptability of the overall residual risk based on the manufacturer's policy for determining acceptable risk.
- Method to verify the implementation and effectiveness of risk control measures.
- Method to collect and review production and post-production information.

A risk plan template is provided at the end of this book.

The risk management file contains all relevant records and documentation applicable to the risk management process. Maintaining a well-organized file facilitates the risk management process and allows for more efficient auditing. Traceability is essential to demonstrate that the risk management process has been applied to each identified hazard. In addition to traceability, completeness is critical. Incomplete tasks within the risk management file may indicate uncontrolled hazards, which could lead to harm. Incompleteness could arise at any stage of the risk management process—whether hazards are unidentified, risks are not assessed, risk control measures are not specified or implemented, or risk control measures prove ineffective.

4.5. Risk Management File

For the particular medical device being considered, the manufacturer shall establish and maintain a risk management file. In addition to the requirements of other clauses of this document, the risk management file shall provide traceability for each identified hazard to:

- The risk analysis;
- The risk evaluation;
- The implementation and verification of the risk control measures; and
- The results of the evaluation of the residual risks.

The risk management file is better defined as a set of records and documents produced by the risk management process. The risk management file generally consists of the following documents:

- Risk policy (establishment of criteria for risk acceptability);
- Risk plan;
- Risk assessment matrix;
- Risk management report;
- Risk benefit report.

The risk management file will also contain the following information:

- Traceability for each identified hazard within the risk analysis;
- Implementation and verification of risk control measures;
- Results of residual risk evaluations;
- Intended use and reasonably foreseeable misuse;
- Qualitative and quantitative characteristics that could affect the safety of the medical device;
- Hazards or events that could result in a hazardous situation;
- Estimation of risk weightings;
- A system for qualitative or quantitative categorization of the probability of harm occurrence and the severity of harm;
- Risk control measures;
- Verification of the implementation of the risk control measures;
- Evaluation of the overall residual risk;
- Risk management report, including a risk review;
- Results from the review of post-production information;
- Decisions arising from the review of post-production information.

Section 5 of the ISO 14971 standard outlines the details behind the risk analysis process. These risk management activities should ideally be recorded in the risk management file. It is crucial to identify which medical device a particular risk management file is applicable to. If a risk analysis is available for a similar device, the identified risks from that device can be used as a starting point when initiating a risk assessment for a new, but similar, device.

The risk analysis should identify both qualitative and quantitative characteristics that could impact the safety of the medical device. The manufacturer must also identify any known and foreseeable hazards associated with the device. These hazards should be based on the intended use, reasonably foreseeable misuse, and safety-related characteristics in both normal and fault conditions. Additionally, for each identified hazard, the manufacturer must consider the combination of events that could result in a hazardous situation. The resulting hazardous situations must be documented in the risk assessment. The differences between hazards, hazardous situations, and harms are further elaborated in this volume.

5. Risk Analysis

5.1. Risk Analysis Process

The manufacturer shall perform a risk analysis for the particular medical device as described in sections 5.2 to 5.5. The implementation of the planned risk analysis activities and the results of the risk analysis shall be recorded in the risk management file.

- *Note 1:* If a risk analysis or other relevant information is available for a similar medical device, that analysis or information can be used as a starting point for the new risk analysis. The degree of relevance depends on the differences between the medical devices and whether these introduce new hazards or significant differences in outputs, characteristics, performance, or results. The extent of use of an existing risk analysis is based on a systematic evaluation of the effects that the differences can have on the occurrence of hazardous situations.
- *2:* See ISO/TR 24971 for guidance on selected risk analysis techniques and on risk analysis techniques for in vitro diagnostic medical devices.

In addition to the records required in sections 5.2 to 5.5, the documentation of the conduct and results of the risk analysis shall include at least the following:

- Identification and description of the medical device that was analyzed;
- Identification of the person(s) and organization who carried out the risk analysis; and
- Scope and date of the risk analysis.

Note 3: The scope of the risk analysis can be very broad (e.g., for the development of a new medical device with which a manufacturer has little or no experience) or limited (e.g., for analyzing the impact of a change to an existing medical device for which much information already exists in the manufacturer's files).

Compliance is checked by inspecting the risk management file.

5.2. Intended Use and Reasonably Foreseeable Misuse

The manufacturer shall document the intended use of the medical device being considered. The intended use should take into account information such as the intended medical indication, patient population, part of the body or type of tissue interacted with, user profile, use environment, and operating principle. The manufacturer shall also document reasonably foreseeable misuse. This documentation shall be maintained in the risk management file.

- *Note 1:* The use specification (see 3.23 of IEC 62366-1:2015) can be an input to determining the intended use.
- *Note 2:* See ISO/TR 24971 for factors to consider in determining the intended use and for an explanation of reasonably foreseeable misuse.

Compliance is checked by inspecting the risk management file.

5.3. Identification of Characteristics Related to Safety

For the particular medical device being considered, the manufacturer shall identify and document those qualitative and quantitative characteristics that could affect the safety of the medical device. Where appropriate, the manufacturer shall define limits for those characteristics. This documentation shall be maintained in the risk management file.

- *Note 1:* See ISO/TR 24971 for a list of questions that can guide in identifying medical device characteristics that could impact safety.
- *Note 2:* Characteristics related to loss or degradation of clinical performance that could result in unacceptable risk are sometimes referred to as essential performance (e.g., IEC 60601-1).

Compliance is checked by inspecting the risk management file.

5.4. Identification of Hazards and Hazardous Situations

The manufacturer shall identify and document known and foreseeable hazards associated with the medical device based on the intended use, reasonably foreseeable misuse, and safety-related characteristics in both normal and fault conditions. For each identified hazard, the manufacturer shall consider the reasonably foreseeable sequences or combinations of events that could result in a hazardous situation and document the resulting hazardous situation(s).

- *Note 1:* A sequence of events can be initiated in all phases of the life cycle, such as during transport, storage, installation, maintenance, routine inspection, decommissioning, and disposal.

- *Note 2:* The relationship between hazard, hazardous situation, and harm, including examples, is explained in Annex C.
- *Note 3:* Risk analysis includes the examination of different sequences or combinations of events related to a single hazard that can lead to different hazardous situations. Each hazardous situation can lead to different types of harm.
- *Note 4:* Systematic techniques for risk analysis that cover specific situations can be used when identifying previously unrecognized hazardous situations. Guidance on some available techniques is provided in ISO/TR 24971.

The documentation shall be maintained in the *risk management file.*

Compliance is checked by inspection of the *risk management file.*

5.5. Risk Estimation

For each identified hazardous situation, the manufacturer shall estimate the associated risks using available information or data. For hazardous situations where the probability of harm cannot be estimated, the possible consequences shall be listed for use in risk evaluation and risk control. The results of these activities shall be recorded in the risk management file.

The system used for qualitative or quantitative categorization of the probability of harm and the severity of harm shall be recorded in the risk management file.

- *Note 1:* Risk estimation incorporates an analysis of the probability of harm occurrence and the severity of the harm. Depending on the application area, only certain elements of the risk estimation process might need detailed consideration. For example, when harm is minimal, an initial hazard and consequence analysis could suffice, or when insufficient information is available, a conservative estimate of probability can give an indication of risk. See also ISO/TR 24971.
- **Note 2:** Risk estimation can be qualitative or quantitative. Methods of risk estimation, including those for systematic faults, are described in ISO/TR 24971, which also provides useful information for estimating risks for in vitro diagnostic medical devices.

- **Note 3:** Information or data for estimating risks can be obtained from sources such as:
  - Published standards.
  - Scientific or technical investigations.
  - Field data from similar medical devices already in use, including publicly available incident reports.
  - Usability tests involving typical users.
  - Clinical evidence.
  - Results of relevant investigations or simulations.
  - Expert opinions.
  - External quality assessment schemes for in vitro diagnostic medical devices.

Compliance is checked by inspecting the risk management file.

Decisions must be made regarding the acceptability of risk. Manufacturers can use the estimated risks and evaluate them based on the criteria for risk acceptability as defined in the risk management plan. They can then assess the risks to determine which ones need to be controlled. The ISO 14971 standard outlines the following guidelines on risk evaluation:

6. Risk Evaluation

For each identified hazardous situation, the manufacturer shall evaluate the estimated risks and determine if the risk is acceptable, using the criteria for risk acceptability as defined in the risk management plan.

If the risk is acceptable, it is not necessary to apply the requirements outlined in sections 7.1 to 7.5 for this hazardous situation (i.e., proceed to section 7.6), and the estimated risk shall be treated as residual risk.

If the risk is not acceptable, the manufacturer shall perform risk control activities as described in sections 7.1 to 7.6.

The results of this risk evaluation shall be documented in the risk management file.

Compliance is checked by inspecting the risk management file.

There are several options available to control risks. The first option should always focus on safe design and manufacture, as design solutions inherent to the medical device's characteristics are likely to remain effective over time. This is crucial because protective measures can fail, and information for safety may not always be followed. While the medical device should ideally be designed and manufactured to be inherently safe, if this is not feasible, then protective measures—such as barriers or alarms—should be implemented. The third option is to provide information for safety, such as written warnings or contraindications. Training for users can also be an important aspect of delivering safety information. The manufacturer may consider providing mandatory training for the intended users.

Additionally, the manufacturing process itself can contribute to risks. For example, contamination of components, residues of hazardous substances used in the process, or mix-ups of parts can all pose risks. These risks can be controlled by designing the manufacturing process to be inherently safe. This could involve eliminating hazardous substances or using separate production lines, or by applying protective measures such as visual inspection steps in the process.

Despite the available risk control options, there may be situations where it is not practical to reduce the risk to acceptable levels according to the pre-established criteria for risk acceptability. In such cases, a benefit-risk analysis can be conducted to determine whether the medical device's benefit to the patient outweighs the residual risk. This option is included at this point in the process to ensure that all possible efforts have been made to reduce risks to the pre-established acceptable levels.

7. Risk Control

7.1. Risk Control Option Analysis

The manufacturer shall determine risk control measures that are appropriate for reducing risks to an acceptable level. The manufacturer shall use one or more of the following risk control options in the priority order listed:

- Inherently safe design and manufacture
- Protective measures in the medical device itself or in the manufacturing process
- Information for safety and, where appropriate, training for users

Ø Note 1*:* The rationale for the priority order in selecting the risk control options is provided in A2.7.1.

Ø Note 2*:* Risk control measures can reduce the severity of harm, the probability of harm occurrence, or both.

Ø Note 3*:* See ISO/TR 24971 for guidance on providing information for safety.

Relevant standards should be applied as part of the risk control option analysis.

Ø Note 4*:* Many standards address inherent safety, protective measures, and information for safety for medical devices. Additionally, some medical device standards have integrated elements of the risk management process (e.g., electromagnetic compatibility, usability, biological evaluation). See ISO/TR 24971 for information on the role of international standards in risk management.

The selected risk control measures shall be documented in the risk management file.

If, during the risk control option analysis, the manufacturer determines that risk reduction is not practicable, a benefit-risk analysis of the residual risk shall be conducted (proceed to section 7.4).

Compliance is checked by inspecting the risk management file.

Two distinct verifications are required when considering the implementation of risk control measures. The first verification ensures that the risk control measure has been implemented in the final design of the medical device or in the manufacturing process. The second verification ensures that the risk control measure effectively reduces the risk.

In some cases, obtaining sufficient data and information for risk estimation can be challenging, which may lead to uncertainty in the residual risk evaluation. In such situations, it may be practical for the manufacturer to focus efforts on verifying the effectiveness of the risk control measures to establish a convincing residual risk evaluation.

7.2. Implementation of Risk Control Measures

The manufacturer shall implement the risk control measures selected in section 7.1. The implementation of each risk control measure shall be verified, and this verification shall be recorded in the risk management file.

Ø **Note 1:** Verification of implementation can be performed as part of design and development verification or process qualification within a quality management system.

The effectiveness of the risk control measures shall also be verified. The results of this verification shall be recorded in the risk management file.

- Ø **Note 2:** Verification of effectiveness can be performed as part of design and development validation within a quality management system and can include testing with users.
- Ø **Note 3:** Verification of effectiveness can also be performed as part of design and development verification or process qualification, if the relationship between the effectiveness in risk reduction and the result of design and development verification or process qualification is known.
- Ø **Examples:**
  - **Example 1:** Design verification of a performance characteristic, such as the dose accuracy of a drug injector, can serve as verification of the effectiveness of risk control measures ensuring safe drug dosing.
  - **Example 2:** Process qualification can serve as verification of the effectiveness of risk control measures related to risks caused by variations in production output.
- Ø **Note 4:** See ISO 13485 for more information on design and development verification and validation. Additional guidance can also be found in ISO/TR 24971.

Compliance is checked by inspecting the risk management file.

Residual risk evaluation is crucial for determining whether the implemented risk control measures have reduced the risk to an acceptable level. If the residual risk exceeds the acceptability criteria established in the risk management plan, the manufacturer must investigate additional risk control measures. This iterative process should continue until either further risk control is not practicable or the residual risk no longer exceeds the acceptability criteria defined in the risk management plan.

7.3. Residual Risk Evaluation

After the implementation of risk control measures, the manufacturer shall evaluate the residual risk using the criteria for risk acceptability defined in the risk management plan. The results of this evaluation shall be recorded in the risk management file.

If the residual risk is not deemed acceptable based on these criteria, further risk control measures shall be considered.

Compliance is checked by inspecting the risk management file.

There may be particular hazardous situations where the risk exceeds the manufacturer's criteria for risk acceptability. In such cases, manufacturers may conduct a thorough evaluation to demonstrate that the benefit of the medical device outweighs the risk.

> 7.4. Benefit-Risk Analysis
>
> If the residual risk is not deemed acceptable based on the criteria established in the risk management plan and further risk control is not practicable, the manufacturer may gather and review data and literature to determine if the benefits of the intended use outweigh the residual risk.
>
> If this evidence does not support the conclusion that the benefits outweigh the residual risk, the manufacturer may consider modifying the medical device or its intended use. Otherwise, the risk remains unacceptable.
>
> If the benefits outweigh the residual risk, proceed to section 7.5. The results of the benefit-risk analysis shall be recorded in the risk management file.
>
> - **Note:** See ISO/TR 24971 for guidance on performing a benefit-risk analysis.
>
> Compliance is checked by inspecting the risk management file.

It is also important for manufacturers to note that risk control measures alone or in combination might introduce a new and sometimes quite different hazard, and that one risk control measure introduced to reduce one risk might actually increase another risk.

> 7.5. Risks Arising from Risk Control Measures
>
> The manufacturer shall review the effects of the risk control measures to determine whether:
>
> - New hazards or hazardous situations are introduced; or
> - The estimated risks for previously identified hazardous situations are affected by the introduction of the risk control measures.
>
> Any new or increased risks shall be managed according to sections 5.5 to 7.4. The results of this review shall be recorded in the risk management file.
>
> Compliance is checked by inspecting the risk management file.

At this point, the risks of all the hazardous situations should have been evaluated. The completeness of risk control step was introduced to ensure that no hazardous situations were left out in the intricacies of a complex risk analysis.

7.6. Completeness of Risk Control

The manufacturer shall review the risk control activities to ensure that all identified hazardous situations have been considered and all risk control activities are completed. The results of this review shall be recorded in the risk management file.

Compliance is checked by inspecting the risk management file.

Evaluating the overall residual risk is crucial, and the manufacturer should assess the remaining residual risk associated with a medical device. This evaluation must consider the contribution of all residual risks in relation to the benefits of the device's intended use. A suitable method and criteria for the acceptability of the overall residual risk must be defined in the risk management plan.

The overall risk management process involves identifying hazards and hazardous situations, evaluating risks, and implementing risk control measures. At some point, the manufacturer must step back and assess the combined impact of all individual residual risks, making a decision on whether to proceed with the medical device. It is possible that the overall residual risk may exceed the manufacturer's criteria for risk acceptability, even though individual residual risks do not. In such cases, the manufacturer must find a method to balance the overall residual risk against the benefits of the device.

This is especially important when determining whether a high-risk but highly beneficial medical device should be marketed. Additionally, the manufacturer is responsible for providing users with relevant information on any significant residual risks that remain so they can make informed decisions regarding the use of the medical device. To achieve this, manufacturers are instructed to include relevant information on residual risks in the accompanying documentation, such as labeling or the Instructions for Use. It is important to note that the manufacturer decides what information and how much should be provided.

Before commercialization, the manufacturer must review the risk management file to ensure that the risk management plan has been appropriately implemented, the overall residual risk is acceptable, and suitable methods are in place to collect any information related to risks during the production and post-production phases of the device.

8. Evaluation of Overall Residual Risk

After all risk control measures have been implemented and verified, the manufacturer shall evaluate the overall residual risk posed by the medical device. This evaluation should consider the contributions of all residual risks in relation to the benefits of the intended use, using the method and criteria for acceptability of the overall residual risk as defined in the risk management plan.

If the overall residual risk is deemed acceptable, the manufacturer shall inform users of significant residual risks and include the necessary information in the accompanying documentation to disclose those residual risks.

- *Note 1:* The rationale for disclosing significant residual risks is provided in section A.2.8 of the standard.
- Note 2: See ISO/TR 24971 for guidance on evaluating overall residual risk and disclosing residual risks.

If the overall residual risk is not deemed acceptable in relation to the benefits of the intended use, the manufacturer may consider implementing additional risk control measures or modifying the medical device or its intended use. Otherwise, the overall residual risk remains unacceptable.

The results of the evaluation of the overall residual risk shall be recorded in the risk management file.

Compliance is checked by inspecting the risk management file and the accompanying documentation.

The **risk management review** is a critical step before the commercial release of the device. During this review, the final results of the risk management process are examined. The **risk management report** contains the findings of this review and is an essential part of the risk management file.

A template for the risk management report is provided in the Annexures below.

The risk management report serves as a high-level document, providing evidence that the manufacturer has ensured the risk management plan has been adequately executed, and that the results confirm the required objectives have been achieved. Depending on the life cycle of the medical device, subsequent reviews may be necessary due to production and post-production activities.

It is imperative that manufacturers establish a suitable system to actively collect and review information related to the device during both production and post-production phases. Suitable methods for collecting and documenting such information must be in place.

9. Risk Management Review

Prior to the release of the medical device for commercial distribution, the manufacturer shall review the execution of the risk management plan. This review shall ensure at least the following:

- The risk management plan has been appropriately implemented.
- The overall residual risk is acceptable.
- Appropriate methods are in place to collect and review information during the production and post-production phases.

The results of this review shall be recorded and maintained as the risk management report, which shall be included in the risk management file.

The responsibility for the review shall be assigned in the risk management plan to individuals with the appropriate authority.

Compliance is checked by inspecting the risk management file.

It is important to remember that risk management does not end when a medical device goes into production. Manufacturers must continuously collect and review production and post-production information and assess its relevance to safety. The information gathered may pertain to new hazards or hazardous situations and can influence risk estimates or the balance between benefit and overall residual risk.

The requirements for production and post-production activities are further elaborated in the latest version of ISO 14971.

10. Production and Post-Production Activities

10.1. General

The manufacturer shall establish, document, and maintain a system to actively collect and review information relevant to the medical device during the production and post-production phases. When establishing this system, the manufacturer shall consider appropriate methods for the collection and processing of information.

Compliance is checked by inspecting the relevant documents.

10.2. Information Collection

The manufacturer shall collect, where applicable:

- Information generated during production and monitoring of the production process;
- Information generated by the user;

- Information generated by those responsible for the installation, use, and maintenance of the medical device;
- Information generated by the supply chain;
- Publicly available information; and
- Information related to the generally acknowledged state of the art.
- **Note:** Information related to the generally acknowledged state of the art can include new or revised standards, published validated data specific to the application of the medical device, the availability of alternative medical devices and/or therapies, and other relevant information (see also ISO/TR 24971).

The manufacturer shall also consider the need to actively collect and review publicly available information about similar medical devices and other similar products on the market.

Compliance is checked by inspecting the relevant documents.

10.3. Information Review

The manufacturer shall review the information collected for its possible relevance to safety, particularly considering whether:

- Previously unrecognized hazards or hazardous situations are present;
- An estimated risk arising from a hazardous situation is no longer acceptable;
- The overall residual risk is no longer acceptable in relation to the benefits of the intended use; or
- The generally acknowledged state of the art has changed.

The results of this review shall be recorded in the risk management file.

Compliance is checked by inspecting the risk management file.

10.4. Actions

If the collected information is determined to be relevant to safety, the following actions apply:

1. **Concerning the Particular Medical Device:**
    - The manufacturer shall review the risk management file and determine if reassessment of risks and/or assessment of new risks is necessary.

- If a residual risk is no longer acceptable, the impact on previously implemented risk control measures shall be evaluated, and this evaluation should be considered when modifying the medical device.
- The manufacturer should consider the need for actions regarding medical devices already on the market.
- All decisions and actions shall be recorded in the risk management file.

2. **Concerning the Risk Management Process:**
   - The manufacturer shall evaluate the impact on previously implemented risk management activities.
   - The results of this evaluation shall be considered as input for top management's review of the suitability of the risk management process.

- *Note*: Some aspects of post-production monitoring are subject to national regulations, and additional measures may be required, such as prospective post-production evaluations.

Compliance is checked by inspecting the risk management file and other relevant documents.

4.20. DIFFERENCE BETWEEN HAZARD, HAZARDOUS SITUATIONS, AND HARMS (AS DEFINED BY ISO 14971)

It is also important to understand the difference between a hazard, hazardous situation, and harm. Annex C of ISO 14971 further explains the differences between this terminology. The risk assessment requires the manufacturer to identify and compile a list of known and foreseeable hazards associated with the medical device. This should be done in both normal and fault conditions. It is important to note that a hazard cannot result in harm until a sequence of events or other circumstances leads to a hazardous situation. At this stage, the risk can be assessed by estimating both the severity and probability of occurrence of harm that could occur (Figure 4.3).

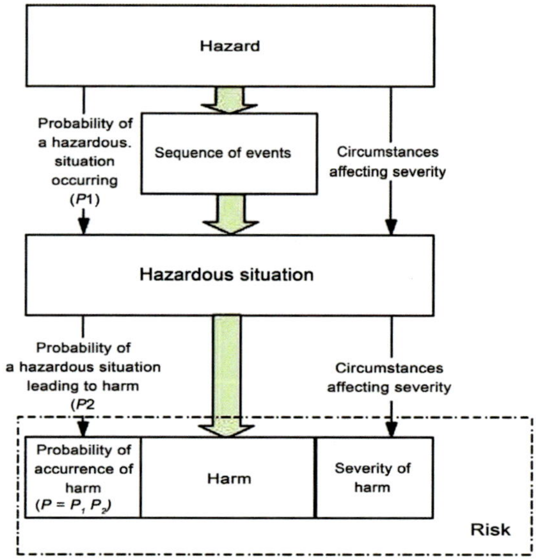

Figure 4.3. *Relationship between hazard, sequence of events, hazardous situation, and harm.*

It is also important to note that, depending on the complexity of the device in question, a hazard can lead to multiple hazardous situations, and each hazardous situation can lead to multiple harms. When determining hazards, hazardous situations, and harms, a good starting point is a review of what has been determined for the same or similar medical devices. This review should take into account a manufacturer's own experience as well as, where applicable, the experience of other manufacturers. This could include information from adverse event databases, publications, scientific literature, and other available sources. This is helpful when identifying and listing typical hazards and related hazardous situations, as well as the associated harm. It is also important to remember that while the list of hazards, hazardous situations, and sequence of events should be completed as early as possible in the design and development process, this is an ongoing activity that continues throughout the life cycle of the device and through to post-production and disposal of the device.

Annex C of ISO 14971 provides a non-exhaustive list of possible hazards that could occur with medical devices, as well as a list of events and circumstances that could result in hazardous situations, which could subsequently result in harm.

Examples of hazards that could be associated with a particular medical device are listed below:

Energy Hazards	**Biological and Chemical Hazards**	**Performance Related Hazards**
Acoustic Energy: o Infrasound o Sound pressure o Ultrasonic o Electric energy o Electric fields o Leakage current o Earth leakage o Enclosure leakage o Magnetic fields o Static discharge o Voltage Mechanical Energy o Kinetic energy o Falling objects o High pressure fluid injection o Moving parts o Vibrating parts • Potential (Stored) Energy o Bending o Compression o Cutting, shearing o Gravitational pull	Biological Agents: o Bacteria o Fungi o Parasites o Prions o Toxins o Viruses • Chemical Agents: o Carcinogenic, mutagenic, reproductive o Caustic, corrosive § acidic § alkaline § oxidants o Flammable, combustible, explosive o Fumes, vapours o Osmotic o Particles (including micro- and nanoparticles) o Pyrogenic o Solvents o Toxic § asbestos § heavy metals § inorganic toxicants § organic toxicants § silica	• Data: o access o availability o confidentiality o transfer o integrity • Delivery: o quantity o rate • Diagnostic Information: o examination result o image artefacts o image orientation o image resolution o patient identity/ information • Functionality o alarm o critical performance o measurement

• o Suspended mass o Tension o Torsion • Radiation Energy o Ionizing radiation § accelerated particles (alpha particles, electrons, protons, neutrons) § gamma § x-ray o Non-ionizing radiation § infrared § laser § microwave § ultraviolet • Thermal Energy o Cryogenic effects o Hyperthermic effects	• • Immunological agents o Allergenic § antiseptic substances § latex o Immunosuppressive o Irritants § cleaning residues o Sensitizing	

- **Examples of Events and Circumstances:**

When identifying foreseeable sequences of events, it is helpful to consider events or circumstances that can cause this sequence of events to occur. The table below provides common examples of events and circumstances.

General Category	Events and Circumstances
Requirements	Inadequate specification of: • Design parameters • Operating parameters • Performance requirements • In-service requirements (e.g., maintenance, reprocessing) • End of life

Manufacturing processes	Insufficient control of: • Manufacturing *processes* • Changes to manufacturing *processes* • Materials • Materials compatibility information • Subcontractors
Transport and storage	• Inadequate packaging • Contamination or deterioration • Inappropriate environmental conditions
Environmental factors	• Physical factors (e.g., heat, pressure, time) • Chemical factors (e.g., corrosion, degradation, contamination) • Electromagnetic fields (e.g., susceptibility to electromagnetic disturbance) • Inadequate supply of power • Inadequate supply of coolant
Cleaning, disinfection, and sterilization	• Lack of validated *procedures* • Inadequate specification of requirements • Inadequate performance of cleaning, disinfection or sterilization
Disposal and scrapping	• No or inadequate information provided *Use error*
Formulation	• Biodegradation • Biocompatibility • No information or inadequate specification provided • Incorrect formulations *Use error*

Usability	• Confusing or missing instructions for use • Complex or confusing control system • Ambiguous or unclear state of the *medical device* • Ambiguous or unclear presentation of settings, measurements or other information • Misrepresentation of results • Insufficient visibility, audibility or tactility • Poor mapping of controls to actions, or of displayed information to actual state • Controversial modes or mapping as compared to existing equipment • Use by unskilled or untrained personnel • Insufficient warning of side effects • Inadequate warning of *hazards* associated with re-use of single-use *medical devices* • Incorrect measurement and other metrological aspects • Incompatibility with consumables, accessories, other *medical devices* • Incorrect patient identification • Slips, lapses and mistakes
Functionality	• Loss of electrical or mechanical integrity • Deterioration in performance (e.g., gradual occlusion of fluid or gas path, change in resistance to flow, electrical conductivity) as result of ageing, wear and repeated use • Failure of a component due to ageing, wear or fatigue
Security	• Unsecured data ports that are externally accessible (e.g., network, serial or USB ports) • Data without encryption • Software vulnerabilities that can be exploited • Software updates without authenticity confirmation

- **Examples of Relationships between Hazards, Foreseeable Sequence of Events, Hazardous Situations and Harms which May Occur:**

The below table highlights the relationships between hazards, foreseeable sequence of events, hazardous situations and harms. Common examples are used.

Hazard	Foreseeable Sequence of Events	Hazardous Situation	Harm
Electromagnetic energy (high voltage)	• Electrode cable unintentionally plugged into power line receptacle	Line voltage appears on electrodes	Serious burns Heart fibrillation
Chemical (volatile solvent, embolus)	• Incomplete removal of volatile solvent used in manufacturing • Solvent residue converts to gas at body temperature	Development of gas embolism (bubbles in the blood stream) during dialysis	Infarct Brain damage
Biological (microbial contamination)	• Inadequate instructions provided for decontaminating re-used anesthesia tubing • Contaminated tubing used during anesthesia	Bacteria released into airway of patient during anesthesia	Bacterial infection
Functionality (no delivery)	• Electrostatically charged patient touches infusion pump • Electrostatic discharge (ESD) causes pump and pump alarms to fail	Failure to deliver insulin to patient with elevated blood glucose level, no warning given	Minor organ damage Decreased consciousness
Functionality (no output)	• Implantable defibrillator battery reaches the end of its useful life • Inappropriately long interval between clinical follow-up visits	Defibrillator cannot deliver shock when an arrhythmia occurs	Death
Measurement (incorrect information)	• Measurement error • No detection by user	Incorrect information reported to clinician, leading to misdiagnosis and/or lack of proper therapy	Progression of disease Serious injury

4.21. THE RELATIONSHIP BETWEEN ISO 14971 AND OTHER STANDARDS

The current version of ISO 14971 was released in December 2019. This version replaced the previous two versions of the standard, which were EN ISO 14971:2007 and EN ISO 2012. For those of us who don't know, the EN version differentiates the version of the standard that is applicable for sale in Europe. The ISO/TR 24971 standard has since been released, which is the standard for Guidance on the application of ISO 14971.

ISO 24971 is a guidance document that was published specifically for ISO 14971. This standard is useful if additional insights and guidance on the application of ISO 14971 are needed.

The standard aims at providing processes for identifying, analyzing, and reducing risks that are linked with the utilization of medical devices. ISO 14971 is one of the most widely recognized standards used to demonstrate conformance in relation to product safety. The standard takes into consideration the entire lifecycle of a medical device through a risk management system. This means that the development, production, and application of the device are all taken into consideration. To demonstrate compliance with ISO 14971:2019, a risk management system needs to be developed, implemented, and maintained over the entirety of the product lifetime. Additionally, all processes and outcomes need to be documented and kept in a risk management file.

4.22. HOW TO IMPLEMENT ISO 14971?

All medical device manufacturers should have procedures for risk management. In fact, manufacturers should have a risk management system in place that consists not only of risk procedures but also a risk policy, risk plan, risk assessment matrix, and risk management reports and risk-benefit reports. Existing procedures should now have been reviewed and updated to align with the 2019 revision of ISO 14971 to ensure they meet the requirements of the risk management standard.

Despite having risk procedures in place, a manufacturer may struggle to achieve buy-in from its employees. Without a company culture that supports risk management, the risk management process will be unable to achieve full implementation. Implementing risk management should ideally start as early as in the design control stage. Too often, manufacturers try to conduct all risk management activities later on, after design controls have been established; however, this tends to limit the ability of risk management to improve design as well as allow for compliance with the requirements of both medical standards:

ISO 14971 and ISO 13485. By integrating risk management into early design decisions, hazardous situations can be considered and evaluated early on in the process. These risks may then be mitigated through design choices if needed. This requirement is also built into ISO 14971 but is also cross-integrated in ISO 13485 directly. Section 7.3.3 of ISO 13485 specifies that outputs of risk management are one of the design and development inputs. This requirement forces manufacturers to conduct risk management processes during the design phases and use information gleaned from that process in the decisions for the design and development of the device.

4.23. UNDERSTANDING ISO 24971: MEDICAL DEVICES – GUIDANCE ON THE APPLICATION OF ISO 14971

The ISO 24971 standard provides guidance to assist manufacturers in the development, implementation, and maintenance of a risk management process for medical devices that meets the requirements of ISO 14971:2019. The ISO 24971 standard describes the approach that manufacturers can use to develop, implement, and maintain a risk management process conforming to ISO 14971.

It is important to note that the clauses and subclauses of the ISO 24971 guidance document have the same structure and numbering as the clauses and subclauses of ISO 14971:2019. This is to aid facilitation of the use of the guidance document when applying the requirements of the standard.

The scope of the ISO 24971 standard is to provide direction when it comes to the development, implementation, and maintenance of a risk management system. This is defined in the guidance document as follows:

1. Scope

This document provides guidance on the development, implementation, and maintenance of a risk management system for medical devices according to ISO 14971:2019.

The risk management process can be part of a quality management system, for example, one that is based on ISO 13485:2016, but this is not required by ISO 14971:2019. Some requirements in ISO 13485:2016 (Clause 7 on product realization and 8.2.1 on feedback during monitoring and measurement) are related to risk management and can be fulfilled by applying ISO 14971:2019. See also the ISO Handbook: ISO 13485:2016 — Medical devices — A practical guide.

Normative references are those which are referred to in the guidance document.

> **2. Normative References**
> The following documents are referred to in the text in such a way that some or all of their content constitutes requirements of this document. For dated references, only the edition cited applies. For undated references, the latest edition of the referenced document (including any amendments) applies.
>
> ISO 14971:2019, Medical devices — Application of risk management to medical devices

The same terms and definitions given in ISO 14971 apply here.

> 3. Terms and Definitions
>
> For the purposes of this document, the terms and definitions given in ISO 14971:2019 apply.
>
> ISO and IEC maintain terminological databases for use in standardization at the following addresses:
>
> - ISO Online browsing platform: available at https://www.iso.org/obp
> - IEC Electropedia: available at http://www.electropedia.org/
>
> **Note**: The defined terms in ISO 14971:2019 are derived as much as possible from ISO/IEC Guide 63:2019, which was developed specifically for the medical device sector.

It is the responsibility of the manufacturer to maintain an ongoing risk management process that runs throughout the lifecycle of the device in question. This risk management system should be documented and maintained.

> **4. General Requirements for Risk Management System**
> **4.1. Risk Management Process**
> ISO 14971:2019 requires that the manufacturer establishes, implements, documents and maintains an ongoing risk management process throughout the life cycle of the medical device. The required elements in this process and the responsibilities of top management are given in ISO 14971:2019 and explained in further detail in this document.

The ISO 24971 further stresses the power which top management has when assigning authorities and responsibilities. The guidance document further stresses that commitment is required at the highest level for risk management activities to be successful.

4.2. Management Responsibilities

4.2.1. Top Management Commitment

Top management has the responsibility to establish and maintain an effective risk management process. It is important to note the emphasis on top management in ISO 14971:2019 Top management has the power to assign authorities and responsibilities, to set priorities and to provide resources within the organization. Commitment at the highest level of the organization is essential for the risk management process to be effective.

If the manufacturer's organization consists of separate entities, for example, business units or divisions, then top management can refer to those individuals who direct and control the entity implementing the risk management process. Each entity can have its own risk management process (and its own quality management system).

Annex C of ISO 24971 provides guidance on how to define the risk policy as well as what elements should be included. Clause 4.2.2. of ISO 24971 states the following:

4.2.2. Policy for Establishing Criteria for Risk Acceptability

ISO 14971:2019 requires top management to define and document the policy for establishing criteria for risk acceptability. Annex C provides detailed guidance on how to define such a policy and which elements should be included, such as applicable regulations, relevant international standards, the generally acknowledged state of the art and known stakeholder concerns. Annex C also explains the relation between the policy and the criteria for risk acceptability and how these criteria are used in risk control and risk evaluation.

The policy can allow specific criteria for each type of medical device (or medical device family). This can depend on the characteristics of the medical device and its intended use (including the intended patient population). ISO 14971:2019 requires that the policy provides guidelines on how to establish the criteria for acceptability of the overall residual risk.

Annex C of ISO 24971 is elaborated on further in this volume.

Section 4.2.3 of the ISO 24971 guidance document deals with the suitability of the risk management process. Th suitability should be reviewed by top management and this should a general, high-level review. Basic checks during the review can include the effectiveness of the procedures, the adequacy of the criteria established and the effectiveness of any feedback loops into the risk assessment process.

4.2.3. Suitability of the Risk Management Process
ISO 14971:2019 requires *top management* to review the suitability of the *risk management process* at planned intervals. The review of the suitability is a high-level review of the *risk management process* and can include reviewing the following aspects, for example:

- the effectiveness of the implemented *risk management procedures*;
- the adequacy of the criteria for *risk* acceptability, which can imply the need for an adaptation of the criteria for *risk* acceptability for specific *medical devices*; and
- the effectiveness of the feedback loop of the production and *post-production* information.

The competence of personnel forms the responsibility of top management. Competence ca n be established through knowledge and/or experience. The ISO 24971 guidance document also provides a guide on what top management should consider when it comes to evaluating the education of risk management experts.

4.3. Competence of Personnel
Ensuring the assignment of competent personnel is a responsibility of top management. Examples of the personnel that can be involved in specific risk management tasks and the relevant knowledge and experience supporting effective completion of the associated tasks are given in Table 4.3.

Some risk management activities can be performed by external consultants or specialists. The required competence should be documented as well as the objective evidence of the fulfilment of these requirements.

Table 4.3. *Examples of Competent Personnel and Relevant Knowledge and Experience*

Personnel or Function	Knowledge and Experience
Risk management owner	Medical device risk management process
Engineer or scientist	Medical device technologies, design, and operating principles
Operations	Manufacturing processes
Supply-chain management	Sources of material and services, including outsourced processes

Medical or clinical expert	Clinical evaluation methodologies and requirements. Use in medical practice, including benefits, hazardous situations and possible
Regulatory affairs	Regulatory requirements pertaining to safety and risk management in countries/regions where the medical device is intended to be marketed
Quality assurance	Quality management systems and quality practices
Packaging, storage, handling and distribution	Hazards and risk control measures in relation to packaging, storage, handling and distribution
Service engineer, biomedical engineer or medical physicist	Hazards and risk control measures in relation to installation, maintenance, repair, calibration, service and support processes and practices
Postproduction	Customer complaints and adverse event reporting, post-market surveillance
Information services	Data mining processes, methodologies for literature search
All individuals involved in the review and approval of the records	Expertise in the functional area for which they are reviewing and approving

Consider the need to include the following topics in the education of risk management experts:

- Management of a risk management program for medical devices;
- Ethics, safety, security and liability;
- Concepts of risk acceptability and benefit-risk analysis;
- Probability and statistics for risk management and reliability;
- Risk management and reliability in design and development;
- Relevant standards and regulations;
- Risk estimation including methods to determine the severity and probability of occurrence of harm;
- Risk assessment methodology;
- Methods for risk control;
- Methods for verifying the effectiveness of risk control measures;
- Methods for analyzing production and post-production information.

The ISO 24971 guidance document provides guidance on the risk management plan and what should ideally be included.

Key points to consider when constructing the risk management plan:

- Scope should be clearly defined;
- Roles and responsibilities should be clearly defined;
- Criteria for risk acceptability should be clearly defined;
- Relevant information sources should be clearly defined;
- All risk management activities should be clearly defined.

Key points to remember:

- The risk plan should be reviewed and updated at a predefined frequency;
- Other items can also be included (other than what is stated in the relevant standards) if it adds value to the business.

4.4. Risk Management Plan

4.4.1. General

The risk management plan describes the scope of the risk management activities, the responsibilities and authorities of those involved, the criteria for risk acceptability, the production and post-production information to be collected and reviewed for the medical device, and all risk management activities that are carried out during the entire product life cycle. The risk management plan can be a separate document, or it can be integrated with other documentation, e.g., quality management system documentation. It can be self-contained or it can reference other documents, such as planning of clinical, biological or usability evaluations or planning of post-production activities.

The risk management plan is a "living document" that will be reviewed and updated throughout the life cycle of the medical device as new information becomes available. The information should be collected on a continuous basis, even after the last medical device is sold and placed on the market. ISO 14971:2019 requires that changes to the risk management plan be recorded in the risk management file.

The extent of planned activities and the level of detail of the risk management plan should be commensurate with the level of risk associated with the medical device. The requirements in ISO 14971:2019 are the minimum requirements for a risk management plan. Manufacturers can include other items such as time-schedule, risk analysis tools, or a rationale for the choice of specific risk acceptability criteria.

4.4.2. Scope of the Risk Management Plan

The scope identifies and describes the medical device and the life cycle phases for which each element of the plan is applicable.

Some of the elements of the risk management plan can apply to the product realization process (design, development and production of the medical device). Other elements can apply to the production and post-production phase (such as installation, use, maintenance, decommissioning and disposal of the medical device)

4.4.3. Assignment of Responsibilities and Authorities

The risk management plan identifies the personnel or functions with responsibility for the execution of specific activities related to risk management. In addition, the risk management plan identifies the individuals with appropriate authority to review and approve risk management decisions and actions. This can entail assignment of personnel familiar with the unique characteristics of the medical device (or medical device family) and their possible relevance to safety. This assignment can be included in a resource allocation matrix defined for the specific life cycle phase and the activities covered in the scope of the plan.

4.4.4. Requirements for Review of Risk Management Activities

The risk management plan details how and when the risk management activities will be reviewed for a specific medical device (or medical device family). This should include the review method, the responsible individuals or functions, who is required to participate in the review, and how the review results are managed. The results of the review of planned risk management activities will be consolidated in the risk management report. The requirements for the review of risk management activities can be part of other quality system review requirements, such as design and development review (see ISO 13485).

4.4.5. Criteria for Risk Acceptability

Criteria for risk acceptability are established according to the manufacturer's policy for determining acceptable risk. This includes criteria for situations where the probability of occurrence of harm cannot be estimated, in which case the criteria for risk acceptability can be based on the severity of harm alone. The criteria can be common for categories of similar medical devices (or medical device families).

It is important to establish the criteria for risk acceptability before starting the risk assessment. Otherwise, the results of the risk assessment could influence the decision when establishing the criteria.

See Annex C for further guidance and examples of criteria that are derived from the policy and applied in risk evaluation.

4.4.6. Method to Evaluate Overall Residual Risk and Criteria for Acceptability

The method to evaluate the overall residual risk and the criteria for its acceptability are derived from the manufacturer's policy for establishing criteria for risk acceptability. ISO 14971:2019 requires that the method and the criteria be stated in the risk management plan for the particular medical device under development. Some inputs for and considerations on the evaluation of overall residual risk are listed in Clause 8.

4.4.7. Verification Activities

The risk management plan specifies how the two verification activities required per 7.2 of ISO 14971:2019 are carried out. The risk management plan can detail the verification activities explicitly or by reference to other plans.

Verification of implementation of risk control measures can be part of design review, approval of specifications, design and development verification in a quality management system, or other verification activities in a quality management system.

Verification of the effectiveness of risk control measures can be part of design and development verification in a quality management system. It can require the collection of clinical data, usability studies, etc., as part of design and development validation in a quality management system.

4.4.8. Activities Related to Collection and Review of Production and Post-Production Information

ISO 14971:2019 requires the manufacturer to establish a system to actively collect and review information about the medical device in the production and post-production phases and to review this information for relevance to safety. Thus, it is important that the risk management plan includes the activities necessary to establish this system. Manufacturers should understand that the information to be collected can be voluminous and comes from many disparate sources. Consequently, robust processes should be used to analyze the information and to identify trends that could otherwise go undiscovered, so that appropriate conclusions and actions can be taken. Statistical techniques should be considered to assist in the processing of the collected data.

> The system to actively collect and review information includes monitoring and receiving feedback such as complaints and adverse event reports. In addition, the system should include active solicitation of feedback from users and collection of other relevant information. The manufacturer should consider the extent of these activities and determine which activities are appropriate for the particular medical device.
>
> For example, limited monitoring might be sufficient for medical devices with a long history of use and well understood risks. For medical devices involving novel treatments (e.g., new intended uses) or innovative technologies and possibly with less understood risks, more elaborate monitoring including post-market clinical follow-up (PMCF) studies could be warranted to understand the issues that can arise in the actual use of the medical device. Further guidance is provided in Clause 10.
>
> The method for collecting production and post-production information can be part of established quality management system processes (see for example, 8.2 of ISO 13485:2016). While a reference to an existing procedure can be sufficient in some cases, any requirements specific to the medical device under consideration should be documented in the risk management plan. Details of the monitoring activities and any planned PMCF studies should also be specified in the risk management plan.
>
> The frequency of review of the collected information should be commensurate with the risk and can also depend on the number of medical devices on the market, the number of incidents reported and the severity of harm reported. The collection and review should continue during the expected lifetime of the medical device.

The risk management file ideally holds all relevant records pertaining to the risk management process. ISO 14971:2019 specifies the particular type of records which the risk management file should hold. The file can be built for each individual medical device or per a medical device family.

> **4.5. Risk Management File**
> ISO 14971:2019 requires the manufacturer to establish and maintain a risk management file, which contains records and other documents created during risk management activities for the medical device throughout its life cycle from initial conception until final decommissioning and disposal. The individual clauses in ISO 14971:2019 specify what records and related documents are to be maintained as part of the risk management file. The risk management file should provide the information necessary for the review of the risk management process at any phase in the medical device's life cycle.

The risk management file can be structured and organized for one type of medical device or for a medical device family. It is important that the risk management records can be assembled in a timely fashion throughout the life cycle of the medical device, as the information could be used during the life cycle to support other activities and decision making, for example, during review of production and post-production information, evaluation of the effect of a change to the medical device, or during audits.

The risk management file is a logical construct. It is not necessary that the risk management file physically contains all the required records and related documents. The records and related documents can be part of files required by other systems such as the manufacturer's quality management system. The records and related documents can exist in any format or media (hard copy, electronic records, etc.).

ISO 14971:2019 requires traceability for each identified hazard to the risk analysis, risk evaluation, implementation and verification of risk control measures, and the evaluation of residual risk. Traceability is a requirement to prove that all identified hazards have been completely addressed in the risk management process. A traceability tool can be used to provide an index to each document in the risk management file providing information on the identified hazard. Such an index can be useful in the management of risk knowledge concerning the identified hazards. This index could be used in later activities such as the evaluation of overall residual risk and the review of production and post-production information. Traceability should be updated as new information becomes available and when the medical device is changed.

See Annex G for guidance on building a risk management file for medical devices that were designed without using ISO 14971:2019.

5. Risk Analysis

5.1. Risk Analysis Process

The risk analysis process consists of the following steps, which are explained in further detail in the next subclauses:

- Description of the intended use of the medical device and reasonably foreseeable misuse;
- Identification of the characteristics of the medical device that are related to safety;
- Identification of hazards and hazardous situations associated with the medical device;
- Estimation of risks for each hazardous situation.

5.2. Intended Use and Reasonably Foreseeable Misuse

The intended use should take into account information such as:

- The intended medical indication, e.g., treatment or diagnosis of type 2 diabetes mellitus, cardiovascular disease, bone fracture, infertility;
- Patient population, e.g., age groups (adults, children, adolescent, elderly), gender (male, female), or disease state;
- Part of the body or type of tissue interacted with, e.g., leg or arm;
- User profile, e.g., patient, lay person, health care provider;
- Use environment, e.g., home, hospital, intensive care unit; and
- Operating principle, e.g., mechanical piston driven syringe, X-ray imaging, MR imaging, subcutaneous drug delivery.

Reasonably foreseeable misuse is defined as use of the medical device in a way not intended by the manufacturer, but which can result from readily predictable human behavior. This can relate to use error (slip, lapse or mistake), intentional acts of misuse, and intentional use of the medical device for other (medical) applications than intended by the manufacturer. Cases of reasonably foreseeable misuse can be identified during design and development by an analysis of simulated use, for example, by applying a usability engineering process, or during the post-production phase by an analysis of actual use. Reasonably foreseeable misuse can be identified throughout the life cycle of a medical device, including iterations of design activities, during which the manufacturer's ability to anticipate potential misuse progressively increases.

The usability engineering process can help to determine whether a particular misuse is reasonably foreseeable or not, for example, by observation during usability testing. The usability test might reveal that users could routinely use the medical device in a manner that is not according to the manufacturer's instructions. This misuse can occur due to poor working culture, inadequate risk perception, limited knowledge of the consequences, or because operating procedures are not clear.

The following example illustrates a case of reasonably foreseeable misuse that was identified and analyzed by application of a usability engineering process. More information on usability engineering can be found in IEC 62366-1 and IEC TR 62366-2.

- **Example:** A single-use medical device is designed to be used only once, but it is reasonably foreseeable that some users might attempt to reuse the medical device. Therefore, warnings against reuse and indications of the possible harm resulting from reuse were included in the accompanying documentation. Application of usability engineering according to IEC 62366-1 demonstrated that this information for safety would be effective, i.e., users would know the correct use and understand the risk of reusing the medical device. However, the usability evaluation also showed that some users are likely to disregard this information and intentionally reuse the medical device. Intentional reuse can be considered abnormal use, which is beyond the scope of the usability engineering process, because the associated risks cannot be controlled in the user interface (see 3.1 and 3.26 of IEC 62366-1:2015). Since this behavior can be considered reasonably foreseeable misuse, the risks from such reuse are analyzed in the risk management process and evaluated against the criteria for risk acceptability according to ISO 14971:2019. It could be necessary to implement risk control measures outside the user interface.

It is imperative that the manufacturer identifies those characteristics (qualitative and quantitative) which could affect safety. Annex A of ISO 24971 provides a good guidance to manufacturers in identifying those characteristics related to safety.

5.3. Identification of Characteristics Related to Safety

It is important to identify the characteristics of the medical device that could affect safety. These characteristics can be qualitative or quantitative and can be bound by certain limits. The questions in Annex A cover many aspects of medical devices and can assist in identifying the characteristics related to safety. For every question, it is indicated which factors should be considered in further detail, with the ultimate goal of identifying all hazards and hazardous situations associated with the medical device. The list of questions in Annex A should not be used as a check list. It can also be helpful to review available information and literature, including adverse event reports, for similar medical devices.

A manufacturer can identify the performance or the functions of the medical device that are necessary to achieve its intended use or that could affect safety, and consider whether any hazardous situations could occur, if any of these functions did not perform properly.

5.4. Identification of Hazards and Hazardous Situations

5.4.1. Hazards

A *hazard* is a potential source of a *harm*. Depending on the specific situation, *hazards* can have different origins/natures. Examples of *hazards* are electricity, moving parts, infectious bacteria, chemicals, gases, sharp edges, high currents, temperature, and ionizing radiation.

Hazards associated with the *medical device* can be deduced from the *intended use* and *reasonably foreseeable misuse* as determined in section 5.2 and the characteristics related to *safety* as determined in section 5.3. Annex C of ISO 14971:2019 provides guidance that can help in identifying *hazards* and sequences of events that can lead to *hazardous situations*. Annex H provides similar guidance for *IVD medical devices*, where incorrect diagnostic information can lead to indirect *risks* to patients.

5.4.2. Hazardous Situations in General

Medical devices only cause *harm* if a sequence of events occurs that results in a *hazardous situation*, which then causes or leads to *harm*. Sequences of events can include a chronological series of causes and effects, as well as combinations of concurrent events. A *hazardous situation* occurs when people, property or the environment are exposed to one or more *hazards*.

Hazardous situations can arise even when there are no faults, i.e., in the normal condition for the *medical device* when it is performing as intended. *Hazardous situations* can be intrinsic aspects of certain therapies. For example, an automated external defibrillator (AED) delivers an electric shock to the patient as part of its normal operation. Similarly, wound cauterization involves the application of high energy to a wound site, and a scalpel has a sharp blade intended to make incisions.

Annex A provides guidance in the form of questions on the characteristics of the *medical device* that could affect *safety*. Those characteristics can help in identifying *hazards* and *hazardous situations*. Annex B provides guidance on several techniques that can support a *risk analysis*. Annex H provides specific guidance on identifying *hazards* and *hazardous situations* for *in vitro diagnostic medical devices*.

5.4.3. Hazardous Situations Resulting from Faults

In cases where a *hazardous situation* only occurs due to a fault, the probability of a fault occurring is not the same as the probability of the occurrence of *harm*. A fault can initiate a sequence of events but does not necessarily result in a *hazardous situation*. A *hazardous situation* does not always result in *harm*.

It is important to understand that there are generally two types of faults that can lead to a *hazardous situation*: random and systematic faults.

5.4.4. Hazardous Situations Resulting from Random Faults

Random faults are typically due to physical or chemical causes such as corrosion, contamination, thermal stress, and wear-out. For many random faults, a numerical value can be given for the probability that the fault will occur. Some examples of random faults are:

- the failure of a part such as an integrated circuit in an electronic assembly;
- the contamination of an IVD reagent leading to incorrect results;
- the presence of an infectious or toxic substance in or on a *medical device.*
- **Note:** A quantitative estimate can only be applied to biological *risks* if sufficient information is known about the *hazard* and the circumstances affecting the probability of the *hazardous situation* occurring, for example, in the use of sterility assurance levels.

5.4.5. Hazardous Situations Resulting from Systematic Faults

A systematic fault can be caused by an error in any activity. It will systematically give rise to a failure when some particular combination of inputs or environmental conditions arises, but will otherwise remain latent.

Errors leading to systematic faults can occur in any part of the *medical device* such as hardware and software in electro-mechanical *medical devices*. Systematic faults in labelling can lead to *use errors* for any *medical device*. These systematic faults can be introduced at any time during a *medical device*'s development, manufacture or maintenance. Some examples of systematic faults are:

- an incorrectly rated fuse fails to prevent a *hazardous situation*: the fuse rating could have been incorrectly specified during design;
- a software database does not provide for the condition of full database: if the database is full, it is not clear what the software will do, with possible consequence that the system will simply replace existing data with new data;
- a fluid, used during the production of a *medical device*, has a boiling point lower than body temperature: residues of the fluid can, in certain circumstances, be introduced into the blood, possibly leading to an embolism;
- the antibody in a hepatitis assay does not detect some variants of the virus;

- inadequately designed environmental control leads to contamination with a toxic substance or an infectious agent;
- the user's manual is written so that if a maintenance routine is performed according to the instructions, the user could be injured (e.g., by a sharp probe).

The accurate estimation of the probability of occurrence of systematic faults is difficult. This is primarily for the following reasons.

- The frequency of systematic faults is laborious to measure. Achieving a reasonable level of confidence in the result will not be possible without extensive data on systematic faults or parameters relevant to *risk control*.
- Consensus does not exist for a method to quantitatively estimate the probability of occurrence of systematic faults.

Because *risk estimation* is difficult in these circumstances, the *manufacturer* should not focus on estimating the *risk* of systematic faults but rather on implementing robust systems to prevent systematic faults which could lead to *hazardous situations* or *harm*.

5.4.6. Hazardous Situations Arising from Security Vulnerabilities

Security in this document includes cybersecurity and data and systems security. Security vulnerabilities can lead to loss of data, disclosure of personal health information, unauthorized access to patient *records*, etc. Such situations can initiate sequences of events, which can ultimately lead to *harm* (patient injury or damage to property). For example:

- Loss of confidentiality can lead to the disclosure of personal health information;
- Loss of integrity can lead to incorrectly represented lab results or malfunction of the *medical device*;
- Loss of availability can prevent the use of critical functionality of a *medical device* or can stop the use of a *medical device* altogether.

See Annex F_for further guidance on security.

5.4.7. Sequences or Combinations of Events

The *hazardous situation* can be the result of a sequence or combinations of independent events. The probability *P*1 of the *hazardous situation* occurring is then given by the product of the probabilities of occurrence of the independent events. A sequence of events can have branches leading to different *hazardous situations* and different events can lead to the same *hazardous situation*.

The example is for an electricity *hazard* and is related to an insulated wire inside a medical electrical device. There is a small probability that the insulation material is degraded and becomes damaged by cracks, and that the cracks lead to an exposed wire. The next possible events are that the user connects and turns on the *medical device*, and that (depending on choices in the user interface) the exposed wire now has line voltage. When the user subsequently opens the protective cover, the *hazardous situation* occurs, namely that the user is exposed to the line voltage of 220 V. The combined probability of this sequence of events is *P*1.

The probability that the user actually touches the exposed wire is estimated to be 0,10. Since the user will always experience a shock from the line voltage, the probability of discomfort is *P*2 = 0,10. The probability of a burn is lower (0,01) and the probability of death is even lower (0,001).

A *hazardous situation* (HS1) can lead to different kinds of *harm* (H1 to H3), ranging from discomfort, to a burn to death. The probability that the *hazardous situation* leads to *harm* can have different values depending on the kind of *harm*, which values are described as *P*2(HS1) H1 through *P*2 (HS1) H3 in Figure 4.1 of the standard. The *severity* of *harm* can be affected by the circumstances of the exposure. For example, the consequences of an electric shock can vary from muscle contractions to burns, heart fibrillation or cardiac arrest, depending on voltage, current, duration of the exposure, and location on the human body.

It is emphasized that several scenarios can be relevant, not only those with the highest *severity* of *harm* or with the highest probability of occurrence of *harm*. Other scenarios can also be relevant. The *manufacturer* should consider what the best manner is to document the *hazardous situation*, describing one or more sequences of events that can lead to this *hazardous situation* and the different kinds of *harm* that can occur.

Information about the *medical devices* on the market can be useful in estimating *risk*. Several approaches are commonly employed to estimate probabilities:

- use of historical design and development data;
- prediction of probabilities using analytical or simulation techniques;
- use of experimental data;
- reliability estimates;
- production and *post-production* information;

- use of expert judgment (an expert in this context can be a person competent on the basis of appropriate education, training, skills and experience; see ISO 13485).

Each of these approaches has strengths and weaknesses. Complementary approaches should be used to increase confidence in the results. Expert judgment should be supplemented with one or more of the other approaches wherever possible. When the other approaches cannot be used or are not sufficient, it might be necessary to rely solely on expert judgment.

5.5. Risk Estimation

5.5.1. General

ISO 14971:2019 requires the *manufacturer* to perform *risk estimation*. Various methods can be used to estimate *risk*. Those methods should examine, for example:

- the circumstances in which a *hazard* is present;
- the sequence of events leading to a *hazardous situation*;
- the probability of a *hazardous situation* occurring;
- the probability of a *hazardous situation* leading to *harm*;
- the nature of the *harm* that could result.

Risk should be expressed in terms that facilitate decision making on *risk* acceptability and the need for *risk control*, for example, using *severity* and probability scales. In order to analyze *risks*, their components, i.e., probability and *severity*, should be analyzed separately.

A *risk* chart such as that shown in Figure 4.2 of the standard shows the distribution of the estimated *risks*, which can be useful for later decision making. The *risks* (*R*1, *R*2, *R*3, …) would be plotted on the chart as they are estimated. *Risk* matrices developed from this figure will be used in examples throughout this clause. This does not imply that this method has general applicability to all *medical devices*; however, it can be useful in many instances. If a *risk* chart or *risk* matrix is used for ranking *risks*, the particular *risk* chart or *risk* matrix and the interpretation used should be justified for that application.

5.5.2. Probability

When sufficient data are available to estimate the probability of occurrence of *harm* with adequate confidence, a quantitative method should be used. Otherwise, a qualitative method based on expert judgment is preferable to a quantitative estimate with high uncertainty. An example of this situation is a new *medical device* where suitable quantitative data are not available until design validation or later when *post-production* data become available. For a qualitative method, the *manufacturer* can describe a series of probability levels with descriptors appropriate for the *medical device*.

Although probability is a continuous variable, a number of discrete levels can be used in practice to simplify the analysis. The *manufacturer* decides how many probability levels are appropriate, based on the expected confidence in the estimates. A larger number of probability levels can be used when estimates are made with greater confidence. At least three levels should be identified to facilitate decision making. The levels can be descriptive and qualitative (e.g., not expected to occur during the lifetime of the *medical device*, likely to occur a few times, likely to occur frequently, etc.) or quantitative. *Manufacturers* should define the levels explicitly, so that there will be no confusion over what falls within each level. A particularly effective way is to assign ranges of non-overlapping numerical values to the discrete levels. An example of three qualitative probability levels is given in Table 4.3 of the standard_and an example of five semi-quantitative probability levels in Table 4.5 of the standard.

The definitions of the probability ranges can be the same or different for different product families. For example, a *manufacturer* can choose to use one set of probability ranges for X-ray equipment, but can have a different set of probability ranges for sterile disposable dressings. Scales for probability can include probability of occurrence of *harm* per use, per device, per hour of use, or within a population, etc.

It is required to document the chosen probability levels or ranges and their descriptors in the *risk management file* for the particular *medical device* (see ISO 14971:2019, 5.5).

There are several factors that are important for estimating the probability of occurrence of *harm*. These include, but are not limited to, the following.

- How often is a particular *medical device* used?
- What is the lifetime of the *medical device*?
- Who makes up the user and patient populations?
- What is the number of users/patients?

- How long and under what circumstances is the user/patient exposed?

Probability estimation encompasses the circumstances and the sequences of events from the occurrence of the initiating event through to the occurrence of the *harm*. The probability *P* of occurrence of *harm* can be decomposed into a probability *P*1 that a *hazardous situation* occurs (i.e., that persons are exposed to the *hazard*) and a probability *P*2 that the *hazardous situation* leads to *harm*. See Figure 4.4 in ISO 14971:2019. A decomposition into *P*1 and *P*2 can be useful to estimate the probability *P* of occurrence of *harm*, but such decomposition is not mandatory.

When the probability of occurrence of *harm* is decomposed into *P*1 and *P*2, it could be the case that one of them can be estimated and the other not. In such cases, a conservative approach can be used by setting the unknown probability equal to 1. Such approach can be useful when the estimated probability is either so low that the resulting *risk* becomes clearly insignificant or negligible, or so high that it is clear the resulting *risk* should be reduced.

5.5.3. Risks for Which Probability Cannot be Estimated

Confidence in a *risk* estimate can be enhanced when a quantitative estimate of the probability of occurrence of *harm* is made on the basis of accurate and reliable data, or when a reasonable qualitative estimate is based on a consensus by qualified experts. However, this is not always achievable. For example, the probabilities of systematic faults, such as those discussed in section 5.4.5, are difficult to estimate. When the accuracy of the probability estimate is in doubt, it is often necessary to establish a broad range for the probability or to determine that it is no worse than some particular value.

Examples where probabilities are difficult to estimate include:

- software failure;
- exceptional misuse situations, such as sabotage or tampering with a *medical device*;
- novel *hazards* that are poorly understood, e.g., imprecise knowledge of the infectivity of the causative agent of Bovine Spongiform Encephalopathy (BSE) prevents quantification of the *risk* of transmission;
- certain toxicological *hazards*, such as genotoxic carcinogens and sensitizing agents, where it might not be possible to determine a threshold of exposure below which toxic effects do not occur.

When the probability of occurrence of *harm* cannot be estimated, it is necessary to evaluate the *risk* on the basis of the *severity* of *harm* alone. The *risk control* measures should focus on preventing the *hazardous situation* entirely or on preventing that the *hazardous situation* leads to *harm*. If this is not possible, the *risk control* measures should focus on reducing the *severity* of the *harm*.

An inverse relationship can be presumed between the rigors of the *processes* used in design, development, manufacturing and maintenance and the probability of some systematic faults being introduced or remaining undetected. The required rigor of these *processes* can be determined by taking account of the *severity* of the consequences of systematic faults and the effectiveness of *risk control* measures external to the *medical device*. The more severe the consequences are and the less effective the external *risk control* measures, the more rigorous these *processes* should be.

5.5.4. Severity

To categorize the *severity* of the potential *harm*, the *manufacturer* should use descriptors appropriate for the *medical device*. *Severity* is, in reality, a continuum; however, in practice, the use of a discrete number of *severity* levels simplifies the analysis. In such cases, the *manufacturer* decides how many categories are appropriate and how they are to be defined. The levels should be descriptive and should not include any element of probability. See the examples in Tables 2 and 4 of the standard.

Severity levels are chosen and justified by the *manufacturer* based on the *harms* that could result for a particular *medical device*. The *severity* levels should be defined with sufficient specificity, so that the correct level of *severity* can be assigned to each *harm* identified in the *risk analysis*.

It is required to document the chosen *severity* levels or ranges and their descriptors in the *risk management file* for the particular *medical device* (see ISO 14971:2019, 5.5).

- **Note:** Terminology used by regulators can be useful in describing the levels of *severity* of *harm*.

5.5.5. Examples

Several approaches can be used for qualitative analysis. A typical approach is to use an N-by-M matrix to describe the *severities* and probabilities of occurrence of *harm* associated with each *hazardous situation*. One carefully defines N levels of probability and M levels of *severity*. Each cell of the matrix represents a subset of the full set of possible *risks*.

A simple example is a 3 × 3 *risk* matrix constructed by using the three *severity* levels of Table 2 of the standard as columns and the three qualitative probability levels of Table 3 as rows. The estimated *risks* (*R*1, *R*2, *R*3, …) are entered into the appropriate cells and the result is shown in Figure 4.3 of the standard. *Manufacturers* should make these definitions as device-specific and explicit as needed to ensure their reproducible use.

		Qualitative severity levels		
		Negligible	Moderate	Significant
Qualitative probability levels	High	R_1	R_2	
	Medium		R_4	R_5, R_6
	Low		R_3	

Example of a qualitative 3 x 3 risk matrix

Table 4.4. *Examples of Three Qualitative Sev*erity Levels

Common terms	Possible description
Significant	Death or loss of function or structure
Moderate	Reversible or minor injury
Negligible	No injury or slight injury

Table 4.5. *Examples of Three Qualitative Probability Levels*

Common Terms	**Possible Description**
High	Likely to happen, often, frequently, always. Likely to happen several times during the lifetime of the *medical device.*
Medium	Can happen, but not frequently. Likely to occur a few times during the lifetime of the *medical device.*
Low	Unlikely to happen, rare, remote. Not likely to occur during the lifetime of the *medical device.*

A more elaborate example is a 5 × 5 *risk* matrix constructed by using the five *severity* levels of Table 4.4 of the standard as columns and the five semi-quantitative probability levels of Table 4.5 of the standard as rows. The estimated *risks* (*R*1, *R*2, *R*3, …) are entered into the appropriate cells and the result is shown in Figure 4.4 of the standard.

4.24. UNDERSTANDING ANNEX A OF ISO 24971: IDENTIFICATION OF HAZARDS AND CHARACTERISTICS RELATED TO SAFETY

The ISO 14971 standard requires the manufacturer to identify those characteristics of the device or device family that could affect *safety*. Consideration of these characteristics is an essential step in identifying the *hazards* associated with the *medical device*. One of the recommended ways is to ask a host of questions pertaining to:

- Manufacturing;
- Intended users;
- Intended use;
- Reasonably foreseeable misuse;
- Disposal of the medical device.

ISO 24971 provides a list of questions in Annex A.2 which can assist manufacturers in identifying all the characteristics of the *medical device* that could affect *safety*. The *manufacturer can* further consult relevant clinical literature, applicable regulations, or the essential principles of *safety* and performance for *medical devices* in ISO 16142-1 or for *in vitro diagnostic medical devices* in ISO 16142-2.

List of questions to help manufacturers to identify characteristics of the medical device that could affect safety (as per ISO 24971):

A.2.1. What is the intended use and how is the medical device to be used?	Factors that should be considered include: • what is the medical device's role in relation to: • diagnosis, prevention, monitoring, treatment or alleviation of disease,

	• diagnosis, monitoring, treatment or alleviation of or compensation for an injury, • investigation, replacement, modification or support of anatomy or a physiological process, or • control of conception? • what are the indications for use (e.g., patient population, user profile, use environment)? • what are the contra-indications? • does the medical device sustain or support life? • is special intervention necessary in the case of failure of the medical device? • can the performance of the medical device be impacted in the event of a security breach (performance degradation or loss of availability)? • can unauthorized access, unauthorized activities, or loss of data affect the medical device safety?
A.2.2. Is the medical device intended to be implanted?	Factors that should be considered include: • the location of implantation, • the characteristics of the patient population, • age, • weight, • physical activity, • the effect of ageing on implant performance, • the expected lifetime of the implant, • the reversibility of the implantation, • whether the implant can be modified or configured while implanted, and • the access connection to perform this modification or configuration (e.g., physical access point or wireless connection to the implanted medical device).

A.2.3. Is the medical device intended to be in contact with the patient or other persons?	Factors that should be considered include: • the nature of the intended contact, i.e., surface contact, invasive contact, or implantation; • the period; and • frequency of contact.
A.2.4. What materials or components are utilized in the medical device or are used with, or are in contact with, the medical device?	Factors that should be considered include: • compatibility with relevant substances; • compatibility with tissues or body fluids; • whether characteristics relevant to safety are known; • is the medical device manufactured utilizing materials of animal origin?
A.2.5. Is energy delivered to or extracted from the patient?	Factors that should be considered include: • the type of energy transferred; • its control, quality, quantity, intensity and duration; • whether energy levels are higher than those currently used for similar medical devices.
A.2.6. Are substances delivered to or extracted from the patient?	Factors that should be considered include: • whether the substance is delivered or extracted; • whether it is a single substance or range of substances; • the maximum and minimum transfer rates and control thereof.
A.2.7. Are biological materials processed by the medical device for subsequent reuse, transfusion or transplantation?	Factors that should be considered include: • the type of process and substance(s) processed (e.g., auto-transfusion, dialysis, blood component or cell therapy processing).

A.2.8. Is the medical device supplied sterile or intended to be sterilized by the user, or are other microbiological controls applicable?	Factors that should be considered include: • whether the medical device is intended for single use or reuse packaging; • shelf-life issues; • limitation on the number of reuse cycles; • method of product sterilization; • the impact of other sterilization methods not intended by the manufacturer.
A.2.9. Is the medical device intended to be routinely cleaned and disinfected by the user?	Factors that should be considered include: • the types of cleaning or disinfecting agents to be used and • any limitations on the number of cleaning cycles. The design of the medical device can influence the effectiveness of routine cleaning and disinfection. In addition, consideration should be given to the effect of cleaning and disinfecting agents on the safety or performance of the medical device.
A.2.10. Does the medical device modify the patient environment?	Factors that should be considered include: • temperature; • humidity; • atmospheric gas composition; • pressure; • light
A.2.11. Are measurements taken?	Factors that should be considered include: • the variables measured and • the accuracy and • the precision of the measurement results • whether the measurement apparatus or data can be compromised. • In addition, the need for calibration and maintenance should be considered.

A.2.12. Is the medical device interpretative?	Factors that should be considered include: • whether conclusions are presented by the medical device from input or acquired data, • the algorithms used, and • confidence limits. Special attention should be given to unintended applications of the data or algorithm, as well as unauthorized manipulation or changes to algorithms and data.
A.2.13. Is the medical device intended for use in conjunction with other medical devices, medicines or other medical technologies?	Factors that should be considered include: • identifying any other medical devices, medicines or other medical technologies that can be involved; • the potential problems associated with interactions (such as the medical device impacting the performance of other medical devices); an • whether the patient follows the instructions for the therapy.
A.2.14. Are there unwanted outputs of energy or substances?	Energy-related factors that should be considered include: • noise and vibration, • heat, • radiation (including ionizing, non-ionizing, and ultraviolet/visible/infrared radiation), • contact temperatures, • leakage currents, and • electric or magnetic fields. Substance-related factors that should be considered include:

	• substances used in manufacturing, cleaning or testing having unwanted physiological effects if they remain in the product. Other substance-related factors that should be considered include: • discharge of chemicals, • waste products, and • body fluids.
A.2.15. Is the medical device susceptible to environmental influences?	Factors that should be considered include: • the operational environment • the transport environment and • the storage environments. These include: • light, • temperature, • humidity, • vibrations, • spillage, • susceptibility to variations in power and • cooling supplies, and • electromagnetic interference.
A.2.16. Does the medical device influence the environment?	Factors that should be considered include: • the effects on power and cooling supplies; • emission of toxic materials; • the generation of electromagnetic disturbance.
A.2.17. Does the medical device require consumables or accessories?	• Factors that should be considered include: specifications for such consumables or accessories and • any restrictions placed upon users in their selection of these.

A.2.18. Is maintenance or calibration necessary?	Factors that should be considered include: • whether maintenance or calibration are to be carried out by the user or by a specialist; • whether special substances or equipment are needed for proper maintenance or calibration; • traceability of the calibrator values to a higher order reference; • how to determine when maintenance or re-calibration is needed; • how to verify that calibration is (still) acceptable.
A.2.19. Does the medical device contain software?	Factors that should be considered include: • whether software is intended to be installed, verified, modified or exchanged by the user or by a specialist, and • the authenticity of a software update.
A.2.20. Does the medical device allow access to information?	Factors that should be considered include: • accessible Ethernet ports, • USB ports, • serial ports, and • removable hard drives.
A.2.21. Does the medical device store data critical to patient care?	Factors that should be considered include: • the possibility of the data being modified or corrupted, • unauthorized access to the data, and • the consequences for the patients.
A.2.22. Does the medical device have a restricted shelf life?	Factors that should be considered include: • whether the medical device can deteriorate over time, • the impact of storage conditions and primary packaging, • the communication of the expiry date (by labelling or an indicator), • possibility of use after the expiry date, and • the disposal of expired medical devices.

A.2.23. Are there any delayed or long-term use effects?	Factors that should be considered include: • ergonomic and cumulative effects. Examples could include: • pumps for saline that corrode over time, • mechanical fatigue, • loosening of straps and attachments, • vibration effects, • labels that wear or fall off, • long-term material degradation.
A.2.24. To what mechanical forces will the medical device be subjected?	Factors that should be considered include: • whether the forces to which the medical device will be subjected are under the control of the user or controlled by interaction with other persons.
A.2.25. What determines the lifetime of the medical device?	Factors that should be considered include: • battery depletion, • deterioration of materials and • failure of components due to ageing, wear, fatigue or repeated use. • The availability of spare parts should be considered as well.
A.2.26. Is the medical device intended for single use?	Factors that should be considered include: • whether the medical device self-destructs after use; • whether it is obvious to the user that the medical device has been used.
A.2.27. Is safe decommissioning or disposal of the medical device necessary?	Factors that should be considered include: • the waste products that are generated during the disposal of the medical device itself, and • the proper sanitization (removal) of all sensitive data on the medical device.

	For example, does it contain hazardous material (e.g., toxic chemical or biological agent), or is the material recyclable? If the medical device stores data, proper handling and security of the stored data should be considered, including data removal and retention.
A.2.28. Does installation or use of the medical device require special training or special skills?	Factors that should be considered include: • the complexity and novelty of the medical device and • the knowledge, skills and • ability of the persons installing, maintaining or using the medical device. This can include: • training, • education, • competence assessment, • certification or qualification.
A.2.29. How will information for safety be provided?	Factors that should be considered include: • whether information will be provided directly to the end user by the manufacturer or will it involve the participation of third parties such as installers, care providers, health care professionals, laboratory directors or pharmacists and whether this will have implications for training; • commissioning and transferring to the end user and whether it is likely/possible that installation can be carried out by people without the necessary skills; • based on the type and expected lifetime of the medical device, whether re-training or re-certification of users or service personnel would be indicated.

A.2.30. Are new manufacturing processes established or introduced?	Factors that should be considered include: • the application of new or innovative technology; and • changes in the scale of production. This can also involve: • changes in contract manufacturing, suppliers and vendors.
A.2.31. Is successful application of the medical device dependent on the usability of the user interface? A.2.31.1. Can the user interface design features contribute to use error?	Factors that should be considered include: • control and indicators; • symbols used; • ergonomic features; • physical design; • layout; • hierarchy of operation; • menus for software-driven medical devices; • visibility of warnings; • audibility of alarms; • standardization of color coding.
A.2.31.2. Is the medical device used in an environment where distractions can cause use error?	Factors that should be considered include: • the consequence of use error; • whether the distractions are commonplace; • whether the user can be disturbed by an infrequent distraction; • whether repetitive stress can reduce the user's awareness or attention.
A.2.31.3. Does the medical device have connecting parts or accessories?	Factors that should be considered include: • the possibility of wrong connections, • similarity to other products' connections, • connection force, • feedback on connection integrity, and • over- and under-tightening.

A.2.31.4. Does the medical device have a control interface?	Factors that should be considered include: • spacing, • coding, • grouping, • mapping, • modes of feedback, • blunders, • slips, • control differentiation, • visibility, • direction of activation or change, • whether the controls are continuous or discrete, and • the reversibility of settings or actions.
A.2.31.5. Does the medical device display information?	Factors that should be considered include: • visibility in various environments, • orientation, • the visual capabilities of the user, • populations and • perspectives, • clarity of the presented information, • units, • color coding, and • the accessibility of critical information.
A.2.31.6. Is the medical device controlled by a menu?	Factors that should be considered include: • complexity and • number of layers, • awareness of state, • location of settings, • navigation method, • number of steps per action, • sequence clarity and • memorization problems, and • importance of control function relative to its accessibility and • the impact of deviating from specified operating procedures.

A.2.31.7. Is the successful use of the medical device dependent on a user's knowledge, skills and abilities?	Factors that should be considered include: • the (intended) users, their mental and physical abilities, skill and training; • the use environment, ergonomic aspects, installation requirements; • the capability of intended users to control or influence the use of the medical device; and • the personal characteristics of intended users that can affect their ability to successfully interact with the medical device.
A.2.31.8. Will the medical device be used by persons with specific needs?	Factors that should be considered include: • users with special characteristics, such as disabled persons, the elderly and children, who might need assistance by another person to enable the use of a medical device; • users having wide-ranging skill levels and differing cultural backgrounds and expectations that could lead to differences in what is considered appropriate application of the medical device.
A.2.31.9. Can the user interface be used to initiate unauthorised actions?	Factors that should be considered include: • whether the user interface allows the user to enter an operation mode with restricted access which increases the possibility of use error and thereby the associated risks, and • whether the user becomes aware of having entered such operation mode.
A.2.32. Does the medical device include an alarm system?	Factors that should be considered are: • the risk of false alarms, • missing alarms, • disconnected alarm systems, • unreliable remote alarm systems, and • the user's ability of understanding how the alarm system works.

A.2.33. In what ways might the medical device be misused (deliberately or not)?	Factors that should be considered are: • incorrect use of connectors, • disabling safety features or alarms, • neglect of manufacturer's recommended maintenance, • unauthorized access to the medical device or to medical device functions.
A.2.34. Is the medical device intended to be mobile or portable?	Factors that should be considered are: • the need for grips, • handles, • wheels or • brakes, and • the need for mechanical stability and • durability.
A.2.35. Does the use of the medical device depend on essential performance?	Factors that should be considered are: • the characteristics of the output of life supporting medical devices or • the operation of an alarm.
A.2.36. Does the medical device have a degree of autonomy?	Factors that should be considered include: • awareness of the user when the medical device with a degree of autonomy generates an error, alarm or failure; • awareness of the user when intervention in an autonomously performed action is required; • the ability of the user to intervene in or to abort an action that is performed autonomously; and • the ability of the user to select and perform proper corrective actions.
A.2.37. Does the medical device produce an output that is used as an input in determining clinical action?	Factors that should be considered include: • whether incorrect or delayed outputs can result in direct or indirect risks to patients, e.g., an incorrect diagnosis resulting in delayed or omitted therapy for a patient.

Annex H of ISO 24971 contains additional information that should be considered when estimating risks from In Vitro Diagnostic (IVD) medical devices. It is important for manufacturers to consider that while these questions and/or lists are helpful, they are neither exhaustive nor representative of all medical devices. For this reason, the manufacturer should add questions or skip questions as is relevant to the device in question. In addition, the manufacturer is advised to consider each question, on its own as well as in relation to other questions.

4.25. UNDERSTANDING ANNEX C OF ISO 24971: RELATION BETWEEN POLICY, CRITERIA FOR RISK ACCEPTABILITY, RISK CONTROL, AND RISK EVALUATION

Annex C describes the relation between the manufacturer's policy for determining acceptable risk as defined by top management and the criteria for risk acceptability which is established based on that policy.

4.25.1. Policy for Establishing Criteria for Risk Acceptability

The risk policy provides a framework for establishing the criteria for risk acceptability. This includes both the criteria for acceptability of individual residual risks and the criteria for acceptability of the overall residual risk. The ISO 14971:2019 standard requires that the policy for establishing the criteria for risk acceptability be documented as part of the quality management system documentation. It does not necessarily need to be part of the risk management file however it makes sense to include it here.

A policy for establishing the criteria for risk acceptability could ideally address the following elements:

- Purpose;
- Scope;
- Factors and considerations for determining acceptable risk;
- Approaches to risk control;
- Requirements for approval and review.

The policy and its elements should then be tailored to fit the specific needs of the manufacturer's organization. Each of the elements is discussed in more detail.

It is important to remember that the purpose of the policy describes the goals of the policy for establishing criteria for risk acceptability. An example is given below:

> ***Example:***
>
> *The purpose of the policy is to provide guidance for establishing the criteria for risk acceptability. These criteria are used in the evaluation of residual risks associated with the medical devices manufactured by [manufacturer's name]. The criteria will ensure that the medical devices have a high level of safety consistent with stakeholder expectations.*

The scope of the policy specifies to whom, where and when the policy applies. An example is given below:

> ***Example:***
>
> This policy applies to all persons involved in establishing, reviewing, updating, and approving the criteria for risk acceptability in risk management plans for medical devices designed, developed and/or manufactured by [manufacturer's name] for commercial distribution.

The following factors and considerations should be taken into account when establishing the criteria for *risk* acceptability

- Applicable regulatory requirements in the regions where the *medical device* is to be marketed
- Relevant international standards for the particular type of *medical device*
- Generally acknowledged *state of the art*
    - This can be determined from international standards, best practices, results of accepted scientific research, publications from authorities, and other information for similar *medical devices* and similar other products
- Validated concerns from stakeholders
    - For example, this could be concerns obtained through direct communication from users, clinicians, patients or regulatory bodies, or through indirect communication via news reports, social media or patient forums.
    - It is also important to consider that the perception and understanding of *risk* acceptability can vary between different groups of stakeholders and can be influenced by their background and the nature of their interest

The approaches to risk control can be defined according to ISO 14971:2019. This can include reducing risk as low as reasonably practicable, reducing risk as low as reasonably achievable, or reducing risk as far as possible without adversely affecting the benefit-risk ratio. Another possible approach to *risk control* can be related to the magnitude of the *risk. This means that risk control* could be omitted for small *risks* below a certain limit. Examples are given below:

> ***Example:***
> Risks are reduced as far as possible without adversely affecting the benefit-risk ratio. Consideration is given to whether technically practicable measures would reduce the risk without impacting the intended use or the benefit of the medical device.

> ***Example:***
>
> Risks related to radiation exposure are reduced to a level as low as reasonably achievable (ALARA), taking account of the technical practicability of the risk control measures.

Requirements for approval and review can also be specified in the policy. This may include who approves and, if needed, how often the policy is reviewed.

> ***Example:***
> The policy for establishing the criteria for risk acceptability is approved by [title/function of top management] and is reviewed at least every [X] years by [name of reviewing body].

4.25.2. Criteria for Risk Acceptability

Criteria for risk acceptability is established based on the manufacturer's policy for determining what is acceptable risk. This is also applicable to the criteria for accepting risks when the probability of occurrence of harm cannot be estimated. In this instance the criteria can be based on the severity of harm alone.

The criteria for risk acceptability are recorded in the risk management plan. Specific criteria can be established for each type of medical device or alternatively, it can be established for each medical device family. This is dependent on its characteristics and intended use.

The criteria for risk acceptability can include combinations of qualitative requirements and quantitative limits for specific properties, preferably based on international standards.

It is also important to note that ISO 14971:2019 requires that the criteria for the acceptability of the overall residual risk of a device be established as well. These criteria can be the same or different from the criteria for acceptability of individual risks. The method to evaluate the overall residual risk and the criteria for its acceptability should be documented in the risk management plan. More details on criteria and methods can be found in Clause 8 of ISO 24971.

4.25.3. Risk Control

Risk control is the *process* in which decisions are made and measures implemented so as to reduce and maintain risks within specified levels. Often this process is guided by the policy for establishing criteria for *risk* acceptability. Two common approaches as discussed in Iso 24971 are discussed below.

4.25.3.1. Possible Approach Number One

One possible approach is to consider how practicable the risk control measures are. Being practicable refers to risk control options that are considered viable or capable of being put into practice.

This should not be confused with practicality (i.e., being practical), which refers to measures that are useful or convenient.

Practicability has two main components, namely:

- Technical practicability; and
- Economic practicability.

Technical practicability refers to the ability to reduce the risk regardless of cost while economic practicability refers to the ability to reduce the risk without making the medical device an unsound economic proposition. This might occur because some risk control measures could make the medical device too expensive and therefore unavailable to the market.

The following are a few examples where technical practicability may be questionable:

- Using risk control measures that diminish the effectiveness of the medical device or which compromise the intended use. This may also have a negative effect on the balance between benefit and risk;
- Overly complex procedures for using the medical device such that the probability of use error is increased or the intended use is compromised. This may have a negative effect on the balance between benefit and risk
- Multiple alarms which could create confusion and restrict operation by the user
- The inclusion of too many warnings or caution labels such that the user is hampered in operating the medical device
- Communicating too many residual risks such that the user has difficulty understanding which ones are really important

Economic practicability, on the other hand, involve making trade-offs between accepting risks and the availability of the device and related treatment.

Multiple factors such as the cost and implications of availability are considered when deciding what is practicable. The benefits for public health and for the society as a whole should be considered however it should also be noted that economic practicability should not be used as a rationale for the acceptance of unnecessary risk.

4.25.3.2. Possible Approach Number Two

A second possible approach to risk control is to consider the magnitude of the residual risk. This can include classifying the risk into one of three categories according to its magnitude:

- **Category 1:** The magnitude of residual risk exceeds the manufacturer's criteria for risk acceptability;
- **Category 2:** The residual risk is so small that it can be regarded as insignificant or negligible. In this case removing it would not lead to a lower overall residual risk); or
- **Category 3:** The residual risk is between the two states specified in a) and b).

The risk policy can then direct whether or not risk reduction efforts should continue for residual risks that have been classified as insignificant or negligible (i.e., category b) before proceeding to risk evaluation.

In this approach the manufacturer may use a semi-quantitative risk chart or risk matrix to support the risk estimation. An example of the risk chart or risk matrix is given in Figure 4.4.

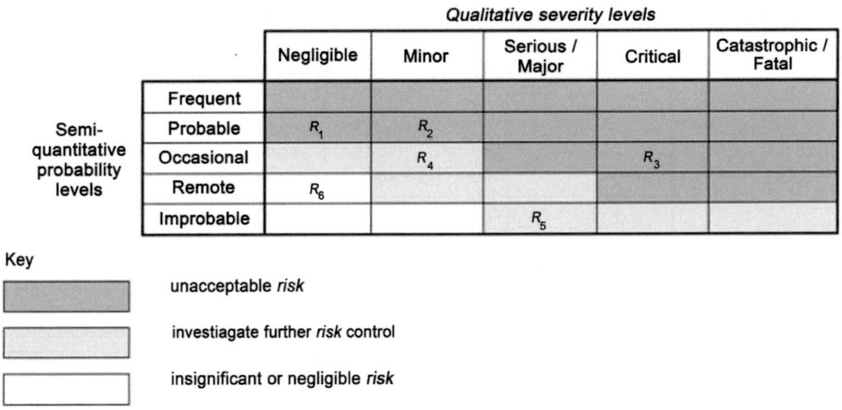

Figure 4.4. *Example of a three-region risk matrix.*

From Figure 4.4, it can be seen that the risk matrix is divided into three regions, namely:

- Unacceptable risk;
- Insignificant or negligible risk; and
- Risks that require investigation to determine if further risk control is feasible.

The estimated risks which are referred too as R1, R2, R3 and so on, have been entered into the appropriate cells. Risks R1 to R3 are not acceptable. Risks R4 and R5 are investigated further, while R6 is insignificant and can be acceptable depending on the manufacturer's policy.

4.25.4. Risk Evaluation

In risk evaluation, the manufacturer compares the estimated risks with the criteria for risk acceptability as defined in the risk management plan and subsequently determines if the residual risks are acceptable or not. A simple risk matrix can support the estimation and evaluation of risk, especially those risks for which no requirements or solutions in international standards exist.

4.25.5. Examples of Applying the Risk Policy

Below are some examples of applying the risk policy to criteria for risk acceptability and how the criteria can be used in risk evaluation. The approach taken by manufacturers can be varied and can include multiple elements and approaches.

Regulatory Requirements	**Policy**	**The Policy Could State:** Criteria meet the safety requirements of the applicable regulations in each market in which the medical device is/will be marketed. For example, regulations require that the medical device maintains safety in single fault condition, including software failures.
	Criteria	The medical devices remain safe in single fault condition, including software failures.
	Evaluation	The medical device is tested and criteria based on testable limits from standards or regulations are applied. Risk evaluation can include inspection of test results, standard conformance reports or certificates.

International standards	Policy	The policy could state: Criteria are based on applicable international product and process standards.
	Criteria	• Testable limits from international product standards are applied. • User interfaces are developed according to the process in IEC 62366-1
	Evaluation	• Inspection of compliance reports for each standard • Inspection of the usability engineering file
State of art	Policy	The policy could state: Criteria are based on the generally acknowledged state of the art, as determined from similar medical devices available on the market and a review of literature on intended use and any alternative therapies or medical devices.
	Criteria	• Leakage currents of the medical device are state of the art, demonstrated by compliance to the limits and tests regarding leakage current of IEC 60601-1. • Dose accuracy of the delivery device are state of the art, as demonstrated by compliance to the limits and tests regarding dose accuracy of ISO 11608-1. • Protection against mechanical failure caused by impact is on the same level as or better than a similar medical device, as demonstrated by comparative test such as drop test.
	Evaluation	Inspection of data and information demonstrating that the medical device conforms to or surpasses the limits based on the state of the art, based on international standards or comparison with a medical device on the market. Risk evaluation can include inspection and comparison of design specifications or comparative test results.

Stakeholder concerns	Policy	The policy could state: Criteria address known stakeholder concerns as identified in a review of medical and scientific literature on the intended use of the medical device, in usability studies, through feedback from advisory boards and/or focus groups, or during post-production monitoring.
	Criteria	• Risks related to bovine materials are a public concern and are essentially eliminated by design. • Risk related to accidental multi-patient use of needle-based medical devices for drug delivery is a concern for clinical organizations, and therefore warnings are required for the risk to be deemed acceptable.
	Evaluation	Risk evaluation can include reviewing performance of the medical device against limits required by the stakeholders, or direct participation of stakeholders (in focus groups or similar) in risk evaluation activities. Risk evaluation can include comparing risk estimations with levels of risk that are considered acceptable by stakeholders.

4.26. UNDERSTANDING ANNEX G OF ISO 24971: COMPONENTS AND DEVICES DESIGNED WITHOUT USING ISO 14971

Annex G of ISO 24071 aims to provide manufacturers with guidance on how available information can be used to be an initial risk management file. The manufacturer can establish the risk management file using available information. Thereafter, the manufacturer could then decide that sufficient evidence is available to demonstrate that the risks for the particular medical device are acceptable, and that the medical device is safe for its intended use; or that additional risk control measures are needed.

Annex G provides further guidance on establishing a risk management plan:

The ISO 14971:2019 standard requires that all risk management activities be planned, especially those activities for the creation of a risk management file to demonstrate that the medical device is safe for its intended use. The mandatory elements of a risk management plan are given in ISO 14971:2019.

The following should be focused on:

- Risk management activities for the remaining phases of the life cycle of the medical device
- The assignment of responsibilities and authorities;
- Requirements for review of risk management activities;
- The criteria for risk acceptability, based on the manufacturer's policy for determining acceptable risk, including criteria for accepting risks when the probability of occurrence of harm cannot be estimated;
- A method to evaluate the overall residual risk and criteria for acceptability of the overall residual risk;
- Verification activities, both for existing risk control measures and for new risk control measures that are considered necessary;
- Activities for the collection and review of production and post-production information, and how this information is used to determine if the risks associated with the medical device are acceptable.

Annex G provides further guidance on establishing a Risk Management File:

A simplified approach to building a risk management file could consist of the following steps:

- Documenting the intended use of the medical device, the reasonably foreseeable misuse and the characteristics related to safety. Reasonably foreseeable misuse can be derived from the information about actual use gathered during the post-production phase.
- Identifying all solutions already adopted in the medical device that can be considered risk control measures.
- Identifying all hazards and hazardous situations associated with the medical device and the possible harm that can result from them.
- Determining if any hazard or hazardous situation exists for which no risk control measure is implemented. In those cases, the manufacturer should estimate and evaluate the risk and apply ISO 14971:2019. For hazards and hazardous situations for which risk control measures are implemented, the manufacturer should verify their effectiveness and estimate and evaluate the residual risk. For residual risks that are not judged acceptable using the criteria for risk acceptability defined in the risk management plan, the manufacturer should consider further risk control and apply ISO 14971:2019.
- Documenting traceability for each identified hazard and hazardous situation to the risk control measures.

- Evaluating the overall residual risk according to ISO 14971:2019 Clause 8.
- Reviewing the execution of the risk management plan according to ISO 14971:2019 Clause 9 and documenting the results in a risk management report.

The records which are generated during these steps form the initial risk management file.

4.27. UNDERSTANDING ANNEX H OF ISO 24971: GUIDANCE FOR *IN VITRO* DIAGNOSTIC MEDICAL DEVICES

4.27.1. Risk Management for IVD Medical Devices

Annex H of ISO 24971 aims to provide guidance for the application of particular aspects of ISO 14971:2019 relating to in vitro diagnostic medical devices. This annex is focused on the indirect risks to patients from incorrect or delayed in vitro diagnostic results. The ISO 24971 standard also addresses risks to device users, other persons and the environment however these are discussed in other areas of the standard.

It is important to also understand that the term "clinician" is used as a general term to mean a healthcare provider who sees patients and who orders, interprets and acts upon IVD examination results.

IVD medical devices and their intended uses are so diverse, it is also important to remember that standards aim to provide general guidance, with the intent to foster critical thinking, cross-functional analysis and informed decision-making within the manufacturer's own risk management process. It is the responsibility of the manufacturer to interpret and apply this information.

The questions and examples in this annex are intended to guide those with appropriate scientific, engineering and clinical expertise to develop and execute effective risk management plans for IVD medical devices. They are not intended to be exhaustive nor necessarily represent best practice for all IVD medical devices. Each manufacturer should determine what is applicable to their particular IVD medical devices.

4.27.2. Context for IVD Risk Management

Managing risks to patients where IVD medical devices are involved can be a challenging process. These risks are often indirect and are characterized by an extended sequences of events that involve trained users who use IVD medical devices to perform IVD examinations and clinicians who rely on the examination results. It is also important for the process between managing risks between manufacturers of IVD medical devices and medical laboratories to talk to each other. For this reason, the ISO 15189 standard and ISO 22367 standard was developed. The ISO 15189 standard is a standard for quality and competence of medical laboratories and requires medical laboratories to control risks to patients. To support this requirement, ISO 22367 was developed to describe a risk management process for medical laboratories based on the same principles and concepts described in ISO 14971:2019. This will promote effective risk communication between manufacturers of IVD medical devices and medical laboratories.

Information for safety as well as the disclosure of residual risks from the manufacturer's risk management process are important inputs to the medical laboratory's risk management process. In addition, the needs of users of IVD medical devices for such information and the laboratory's feedback from using the IVD medical devices are important inputs to the manufacturer's risk management process. Manufacturer must include the user needs for risk management information as design inputs when developing or modifying an IVD medical device.

When a manufacturer supplies an IVD medical device to a medical laboratory, any risks that could not be controlled through design or protective measures are transferred to the laboratory along with the information for safety to control those risks. The manufacturer also discloses any residual risks in the accompanying documentation, so that the laboratory director can evaluate these risks and determine their acceptability.

Manufacturers can provide information for safety to inform users of IVD medical devices, but they cannot influence the actions of clinicians who order, receive and act upon the examination results.

Some IVD medical devices are intended for use by clinicians at the point of care, while self-testing IVD medical devices are actually used by patients. Although similar risk scenarios can exist for these devices, the user's ability to control the risks can be more limited. Therefore, it is important that point of care devices and self-testing devices are designed with risk control measures appropriate for the (intended) users and the (intended) use environment outside laboratories.

4.28. PERFORMING A RISK ANALYSIS OF IVD MEDICAL DEVICES

4.28.1. Intended Use and Reasonably Foreseeable Misuse

H.2.1.1. Analytical and Clinical Use:

Most IVD medical devices have two users.

- a user who performs all or part of an examination – this is referred to as "analytical use" and,
- a clinician who receives, interprets and acts on the examination results – this is referred to as "clinical use".

In the case of IVD medical devices intended for self-testing, the patient can be the only user.

H.2.1.2. Device Description:

Each risk analysis begins with identifying and documenting a clear description of the IVD medical device and its specific role in producing the examination result.

Questions to consider when documenting an IVD medical device description include:

- Is the device used alone to produce examination results or in combination with other devices?
    - o If the device is a standalone analytical system, is it automated (i.e., is software used? Are robotics involved)?
    - o If used in combination with other IVD medical devices to form a system, what is its role in producing the examination result (e.g., is it a sample collection system? is it a sample receptacle? is it a measuring instrument? is it a software? is it a database? is it a reagent? is it a calibrator? is it a control material? is it an accessory? etc.)?
    - o If part of a system, how does the IVD medical device interact with other components of the system?
- Are other reagents or accessories necessary? Are these provided?
- Does the device employ new or novel technology
- Does the device employ digital information technology for documenting and/or transmitting examination results to clinicians or communicating with mobile applications?

- Do software applications provide diagnostic or treatment recommendations?
- Does the IVD medical device communicate with a medical device that immediately administers treatment based on the IVD results? (e.g., is it an IVD medical device that measures blood glucose levels and communicates with an implanted insulin administration system)?

H.2.1.3. Analytical Use:

The intended use of the IVD medical device includes:

- The analyte/analytes intended to be detected or measured;
- Acceptable sample types;
- Calibration;
- Quality control;
- Preventive maintenance activities; and
- The use environment.

It is important that reasonably foreseeable misuse is also considered when determining and managing risks.

Questions to consider when identifying the analytical use of the IVD medical device include:

- What analyte is the device intended to measure or examine?
- Will the examination results be qualitative, semi-quantitative or quantitative?
- Will the device be used in the pre-examination, examination or post-examination phase?
- What specimens can be analyzed (e.g., serum, plasma, blood, urine, other body fluids, tissues)?
- Do other substances potentially found in these samples interfere with the analytical process?
- In nucleic acid sequencing procedures, is the amplicon sensitive to contamination from environmental sources of DNA/RNA?
- Are there any additional limitations for use in specific use environments (These environments may include:
    - o Medical laboratories;
    - o Emergency rooms;
    - o Operating rooms;

- Ambulances;
- Intensive care units;
- Neonatal care units;
- Nursing homes;
- Physician's offices;
- Screening clinics;
- Patient's homes.

- Does the IVD medical device interface, connect or communicate with other devices or networks?
- Who will be using the IVD medical device to perform examinations, and what training and qualifications will be appropriate?

H.2.1.4. Clinical Use:

The Indications for Use, also known as the intended clinical use of the IVD medical device includes the medical conditions and patient populations for which the examination results are used.

The internal or external clinical experts will generally understand the following:

- how the IVD examination results will be used in clinical decision making;
- the medical decision points and degree of accuracy required;
- whether clinicians can recognize incorrect results (This could be based on magnitude of error or consistency with other clinical information);
- what actions the clinician would take in the event of an abnormal or unexpected result;
- the clinical significance of delayed results, if any;
- potential adverse consequences of unnecessary medical intervention

Additional questions to consider when identifying the clinical use include:

- Will the examination results be used for:
    - diagnosis in order to cure, treat or prevent a disease or other condition?
    - measuring body fluid constituents to determine a patient's state of health?
    - monitoring therapeutic drug levels to ensure an effective dose?
    - determining the safety of donated blood or organs?

   - screening a population for the presence or absence of a specific marker?
   - predicting the effectiveness of a therapeutic alternatives?
   - predicting the risk of developing a medical condition?
   - applications other than the intended use?
- What injury, illness or condition will the results be used to detect, diagnose, predict or monitor?
- Who will use the IVD examination results: medical specialists, general clinicians or patients?
- Is the role of the examination results in medical decisions to be used
   - as the basis for immediate medical decisions?
   - with other relevant information to guide a medical decision?
- Which patient populations will primarily experience the benefit from the IVD examinations?
- Should any patient populations be explicitly contraindicated?

H.2.2. Characteristics Related to Patient Safety:

H.2.2.1. General Considerations:

It is important to remember that in addition to biological, chemical, electrical, mechanical and security characteristics which IVD medical devices may have in common with other medical devices, they also have analytical performance and reliability characteristics that determine the suitability for their intended clinical use.

Some IVD medical devices can perform multiple examinations simultaneously. Their clinical performance is based on the interpretation of patterns of results. A common example is multiplex assays. It is also important to remember that some IVD medical devices utilise digital information technology and can have characteristics related to their ability to store and transmit an examination result or ancillary information which may be needed for a medical decision. Failure to meet a performance, reliability or communication requirement can initiate a sequence of events that might result in harm to a patient.

H.2.2.2. Performance Characteristics Related to Patient Safety:

1. **Quantitative Examinations:** These measure a quantity in a representative specimen that has been taken from a patient. The

results are usually expressed as a concentration or percentage. Depending on the medical application false high, false normal or false low results can occur and can potentially affect a diagnosis, thereby causing inappropriate or delayed therapy which could in turn lead to patient harm.

The type and severity of harm will depend on the magnitude of error at certain medical decision points. The relevant performance characteristics of quantitative IVD medical devices may include the following:

- Trueness of the measured values:
    - o Bias;
    - o Traceability to a reference standard.
- Measurement precision:
    - o Repeatability;
    - o Intermediate precision;
    - o Reproducibility.
- Analytical specificity:
    - o Influence of interfering or cross-reacting substances.
- Analytical sensitivity:
    - o Ability to discriminate between quantity limits or ranges.
- Detection limit:
    - o Lowest quantity that can be reliably detected.
- Quantitation limit:
    - o Lowest quantity that can be accurately measured.
- Measuring interval:
    - o Range of values over which the analytical performance was validated.

2. **Semi-Quantitative Examinations:** These provide a clinically useful approximation of the quantity being measured. Values are usually assigned based on an ordinal scale or are reported as a quantity limit. These can be expressed numerically, for example:

- within a specified range of values
- or greater or less than a specific quantity, titre or serial dilution or
- relatively (e.g., as +3, +2, +1 or trace amount).

Common examples of semi-quantitative examinations include:

- urine “dipsticks

- tablets that detect the presence of ketones, and
- serological agglutination *procedures*.

The performance characteristics of semi-quantitative *IVD medical devices* can include:

- analytical sensitivity (ability to discriminate between quantity limits or ranges);
- analytical specificity (influence of interfering or cross-reacting substances)
- detection limit (lowest quantity that can be reliably detected);
- precision of the measured signal values (repeatability, reproducibility).

3. **Qualitative Examinations:** These determine the presence or absence of an analyte. Results are reported as positive, negative or indeterminate. Positive or negative results can be defined by cut-off values and relevant databases.

A positive result is when the analyte is absent or a negative result is when the analyte is present and this may affect the diagnosis or treatment.

The performance characteristics of qualitative *IVD medical devices* can include:

- Analytical sensitivity (fraction of true positive results in samples containing the analyte);
- Analytical specificity (fraction of true negative results in samples containing the analyte);
- Diagnostic sensitivity (fraction of true positive results in patients with disease);
- Diagnostic specificity (fraction of true negative results in patients without disease).

H.2.2.3. Reliability Characteristics Related to Patient Safety:

When clinicians depend on IVD examination results for urgent medical decisions, timely results can be as important as accurate results. Failure to produce a result when it is needed can delay necessary medical intervention.

The reliability characteristics of *IVD medical devices* can include:

- system reliability
    - o mean time between failures
    - o mean time to failure

- component compatibility (including versions and critical tolerances);
- software reliability (error-free operation);
- reagent or control stability;
- system usability (avoidance of *use errors*)

H.2.2.4. Digital Information Technology Characteristics Related to Patient Safety:

Correct identification of the patient and the sample is imperative to suitable operation of the IVD device. It is important to note that some examinations also require ancillary information about the patient, the sample, or the examination for proper interpretation of the results.

If an *IVD medical device* is designed to collect, store and report such information with the examination result, device characteristics leading to data corruption or alteration can contribute to misdiagnosis or inappropriate therapy.

Ancillary patient information required by the clinicians can include:

- Correct patient name and sample identification;
- Patient details such as:
    - Age;
    - Gender;
    - Population;
    - Genetic factors;
    - Medications;
    - Nutritional state.
- Sample details:
    - Sample type;
    - Description;
    - Acquisition time.
- Measurement details:
    - Measurement procedure;
    - Units of measure;
    - Measurement uncertainty.
- Application details:
    - Cut-off points;
    - Reference intervals.

Digital information technology characteristics that can affect patient *safety* include:

- Connections between devices and/or networks (wireless or wired);
- Internet data transmission;
- Interface with digital applications (networked or mobile);
- Applications that emulate results from an *IVD medical device*;
- Embedded software applications (e.g., interpretation or treatment recommendations);
- Unshielded data transfer (e.g., ESD susceptibility);
- Digital data storage (e.g., susceptibility to corruption, manipulation or deletion);
- Disruption of other connected devices (creating additional *hazards*).

H.2.3. Known and Foreseeable Hazards to Patients:

H.2.3.1. Identification of Hazards:

From the viewpoint of a patient, an IVD examination result would be considered a hazard if it could lead to:

- Inappropriate medical interventions that could result in harm; or
- Lack of medical interventions necessary to prevent being harmed.

The following general hazards could cause or contribute to potentially harmful medical decisions. Specific hazards should be identified in terms of the magnitude and direction of error, the extent of delay, or the ancillary information that is incorrect or missing.

In addition to hazard identification for the IVD medical device itself, hazard identification related to connectivity should also be evaluated. The increased use of IVD medical devices connected to other devices or systems, either directly or through a computer network, wireless technology or the internet, has created new challenges for their safe operation.

The need to ensure effective IVD medical device functionality and safety has become more important with the increasing use of connected devices, and the frequent electronic exchange of health information produced by IVD medical devices. The risk management process should be able to identify failures that could cause hazards as described below. Some hazards include:

- Incorrect examination result;
- Delayed examination result;

- Incorrect information accompanying the result.

Let's look at these hazards in more detail:

1. **Incorrect Examination Result:** For quantitative and semi-quantitative examinations, results are considered incorrect if the difference from a correct value exceeds the error limit required for the clinical application.

Such analytical performance requirements are generally established during the design input process.

Some medical decisions can be influenced by the magnitude of the examination result. For this reason, the clinical significance of an incorrect result could depend on the magnitude of the difference between the measured value and the true value.

For qualitative examination procedures such as HIV and pregnancy examinations, in which only a positive or negative result is provided, examination results are either correct, incorrect or indeterminate.

2. **Delayed Examination Result:** A delayed examination result occurs if it is needed for a medical decision and the clinician does not receive it in time to support a critical therapeutic or intervention decision. Criteria can be established to define what constitutes a clinically significant delay for the medical application. An example of this would be an urgent care situation.
3. **Incorrect Information Accompanying the Result:** The consequences of an error in the ancillary information provided with an IVD examination result depends on how the information is used in clinical decision making, and whether the error could cause or contribute to *harm*.

H.2.3.2. Identification of Hazards from Fault Conditions:

IVD medical devices that fail during use can lead to one or more of the general hazards as defined below:

- Incorrect examination results;
- Delayed examination results;
- Incorrect information accompanying the result.

Fault conditions potentially leading to hazards can include the following:

- Within-batch or batch-to-batch inconsistency. Examples include:
    - o Reagents;
    - o Calibrators;

  - Controls.
- Non-traceable value assignments. Examples include:
  - Calibrators;
  - Proficiency materials;
  - Assayed controls.
- Reagent non-specificity. Examples include:
  - Interfering factors;
  - Antibodies.
- Sample or reagent carryover. Example include: pipetting instruments.
- Measurement imprecision. Example include: system-level.
- Unstable materials. Examples include:
  - During transportation;
  - Storage;
  - Use.
- System malfunctions. Examples include:
  - Hardware;
  - Software;
  - Components;
  - Accessories.
- Digital technology failures. Examples include:
  - Software and/or firmware vulnerability to intrusion. For example: data modification or theft
  - Data transfers resulting in incorrect or missing results, inappropriate treatment recommendations, or delays from loss of function due to environmental conditions;
  - Connections disrupting the performance of the connected medical device, creating unsafe conditions for the patient;
  - Digital applications incorrectly connected to another device or digital application;
  - Corruption during data storage that causes incorrect information or delayed results; or
  - Delays in availability of results or patient information due to loss of network connectivity.

When the IVD medical device is used with digital software applications, failures leading to a delay of results may include:

- smart device operating system changes, resulting in application not being available and causing delay of treatment, or in unexpected behavior causing incorrect recommendation for treatment;
- smart device data storage capacity or rate of transfer data limitations, resulting in delay of treatment or incorrect recommended treatment;
- time inconsistencies between application and smart devices, resulting in delay of treatment or incorrect results (specifically related to out-of-date results appearing as valid).

H.2.3.3. Identification of Hazards from Normal Use

The inherent limitations in IVD medical device technology can occasionally lead to one or more of the general hazards to patients even though all warnings, precautions and instructions for use were followed. This may also occur despite the device functioning as intended, and the analytical performance of the device meeting the claims of the manufacturer.

It should be remembered that every examination result is subject to unavoidable sources of variability. Even when the analytical performance has been optimized to minimize the risks, an occasional result in normal use can be a hazard for an individual patient.

Hazards potentially occurring in normal use can include inaccurate results due to the following:

- inherent false negative and false positive rates of qualitative examination procedures. This may be caused by the uncertainty of statistically assigned cut-off values
- measurement uncertainty associated with quantitative examination procedures
- misclassification of results as “abnormal” or outside a “normal” reference interval
- influence of interfering substances in the sample. Common examples include:
  - cross-reacting antibodies
  - certain drugs or biochemical metabolites, or
  - sample preparation materials
- biological variability of the analyte itself. Common examples include:

  - heterogeneity of natural proteins
  - population differences in normal analyte concentrations
- chemical properties of the analyte itself (e.g., intrinsic instability, adhesiveness);
- variability of the patient sample matrix
- the finite reliability of instrument components.

H.2.3.4. Identification of Hazards from Use Errors:

Use errors can cause one or more of the abovementioned general hazards. Non-routine laboratory processes can be especially prone to use error. Reasonably foreseeable use errors (i.e., resulting from readily predictable human behavior) can be identified and potentially prevented by a usability engineering process during IVD medical device design and development.

Use errors potentially leading to IVD hazards in the medical laboratory or at the point of care can include the following:

- Overlooking special requirements (e.g., outside the normal laboratory routine);
- Performing operations out of sequence, including pre-examination and post-examination processes (e.g., unclear instructions, confusing user interface);
- Data entry errors (e.g., patient name, identification number, birth date or age, gender, etc.).

Use errors by patients performing self-testing can include the following:

- applying insufficient volume of sample (e.g., too little for accurate measurement);
- inserting reagent module improperly (e.g., before device is ready for measuring).

H.2.3.5. Identification of Hazards from Reasonably Foreseeable Misuse

A usability engineering process can also help manufacturers of IVD medical devices to prevent hazards and hazardous situations arising from reasonably foreseeable misuse.

Examples of reasonably foreseeable misuse include the following:

- use of an IVD medical device prior to reading the instruction manual or completing training;

- disregard of warnings, instructions, or other information for safety;
- collection of an inappropriate sample types
- reporting examination results for contraindicated or unvalidated clinical use;
- using a self-testing IVD medical device in a critical care setting
- using unverified third-party calibrator, reagent, control material or accessory;
- storing materials in incorrect conditions
- operation of an IVD instrument outside specified environmental conditions;
- disabling, overriding, or failing to enable safety features
- neglecting to perform prescribed instrument maintenance;
- connection to an information system without adequate network connectivity or security;
- malicious intent to create incorrect results or delay in treatment, including:
- hijacked and impersonated device by third-party application or individual to alter results, producing incorrect results on connected digital applications;
- corrupted device software configuration, producing incorrect results;
- intercepted data in transit to delay results or send incorrect results to the user.

Reasonably foreseeable misuse by patients performing self-testing can include the following:

- dividing or reusing reagent test strips (This could be done to reduce cost)
- taking samples from an alternative site (Alternative sites may be used due to pain)
- failing to clean and disinfect the venipuncture site (potential for contamination/infection);
- storing reagent strips in inappropriate environmental condition (an environment which results in overheating).

H.2.4. Identification of Potential Harms:

The ISO 14971:2019 standard requires manufacturers to estimate the risks associated with each identified hazardous situation. This is based on the probability of occurrence and the severity of possible harm. This requires the manufacturer to identify the potential harms to patients with sufficient specificity to assign appropriate severity values.

It is also important to note that for some examinations, a single hazardous situation can result in several different harms representing a range of severities. Manufacturers should determine which harms are appropriate to include in the risk analysis to ensure a high degree of protection of health and safety. Any rationales behind this should also be documented.

Questions that might help to identify and classifying potential harms include:

- Is the intended use a major determinant of therapy for a serious medical condition?
- If so, what harms might occur from a misdiagnosis or inappropriate therapy?
- Does the intended use involve detection of an infectious disease agent
- If so, could a false negative result allow the infection to spread to others in the population?
- Is the intended use to detect and diagnose an inherited condition
- If so, could a false negative result allow progression of an otherwise preventable or treatable disease?
- Could a false positive result lead to unnecessary medical intervention and potential harm?
- Is the intended use to predict drug or device effectiveness?
- If so, could a false negative result cause the loss of therapeutic benefits and subsequent harm?
- Could a false positive result have harmful consequences?
- Is the intended use to screen transfusion or transplant donors?
- If so, could incorrect results cause transmission of disease to recipients or lead to rejection of a properly functioning organ?
- Is the intended use to monitor a critical body function?
- If so, what harms might occur from an incorrect result or a significant delay in receiving the result?
- If medical intervention occurred, would the outcome be irreversible or would the outcome be reversible (with or without further medical intervention)?

- Does the IVD medical device require connection to a network or the internet, where modification or theft of a patient's data could occur (e.g., inadequate security)?

H.2.5. Identification of Hazardous Situations:

The ISO 14971:2019 standard requires *manufacturers* to compile a comprehensive set of *hazardous situations* for the *risk analysis. It also* leaves it up to the *manufacturer* to determine what constitutes a *hazardous situation* for the purpose of the *risk analysis*. One such approach is to review the sequence of events.

Examples of *hazardous situations* for *IVD medical devices* can include events such as:

- Receipt of an incorrect laboratory result by a clinician;
- Delay in therapy (e.g., due to failure of the *IVD medical device*);
- Delay in reporting an urgent laboratory result to a clinician;
- Inappropriate therapy (e.g., based on incorrect self-testing result);
- Misidentification of a patient's sample (e.g., due to *use error*);
- Reporting incorrect information with a patient's result (e.g., due to networking failure).

The following questions can be helpful to analyze *hazardous situations* related to incorrect outcomes:

- Is it possible that a false negative or false positive result would still appear "believable" and not be further confirmed before deciding on a course of clinical action/inaction?
- Are there few, limited or no other diagnostic tools available to confirm or deny a potentially false positive or false negative IVD test result?

H.2.6. Identification of Foreseeable Sequences of Events:

H.2.6.1. General Considerations:

The identification and analysis of reasonably foreseeable sequences or combinations of events that could lead to a hazardous situation and potentially progress to harm is necessary to estimate the probability that the harm would occur. Understanding these activities and events can also help the manufacturer select the hazardous situation for risk analysis and identify opportunities for risk reduction and risk control.

The specific sequence of events to be analyzed will depend on the particular IVD medical device and its intended use. When outlining the sequence of events, the manufacturer should consider the following:

- knowledge of the intended user
- skills of the intended user
- abilities of the intended users
- the use environment
- any events and circumstances that could increase or decrease the risks

Although medical laboratories operate with control procedures designed to detect errors, the manufacturer should consider their effectiveness in detecting specific failure modes.

Sporadic random failures are especially difficult for a laboratory to detect. Clinicians who are experience normally would question any results that are inconsistent with other available information.

H.2.6.2. Description of the Sequence of Events

The risk scenarios for IVD medical devices usually involve a longer sequence of events than for other medical devices. Often it includes activities that are not under the direct control of the manufacturer. In these cases, the description of the sequence of events should start with the initiating event and the events directly under the control of the manufacturer. From here, it should progress logically through the foreseeable decisions and actions of the device users, and end with the clinical decisions and actions leading to each of the harms identified as foreseeable outcomes.

The description should be as detailed as necessary to capture the main initiating and contributing events. However, a balance is needed such that the description is not too detailed such that it hinders the risk analysis.

1. **Events under Control of the Manufacturer:** Failures in the manufacturer's quality management system can result in initiating events which lead to a hazardous situation.

A fault condition or potential for use error can be caused by activities under a manufacturer's direct control. These include:

- design and development
- device labelling
- manufacturing and supplier management
- product inventory and distribution

- equipment servicing
- training and product support

2. **Events under Control of the User of the IVD Medical Device:** During device operation, use errors and device failures could happen. This could occur during device operation in the laboratory or at the point of care. The activities taken to prevent or detect hazards as well as the actions taken in response by the user are under the direct control of the laboratory or point of care facility. These device users should build in protective measures and information for safety as provided by the manufacturer into their own risk management system.

The users of IVD medical devices can also cause risks or contribute to risks to patients through misuse of the device in question. This may include failure to maintain adequate quality assurance procedures, contingency and recovery plans, or security protection. It is also important to remember that decisions to report or not report an examination result to a clinician are completely under the control of the medical laboratory or other users of the IVD medical device.

Capabilities of the intended users and the use environments should be considered when analyzing the sequence of events under control of users of the device in question. Examples of activities under the control of users include:

- selection of sample collection device as well as usage and storage of sample collection device used with the IVD medical device;
- collection, processing and storage of patient samples;
- system installation and setup
- User training, component or consumable qualification, and performance verification;
- selection, preparation, usage and storage of accessories
- selection, preparation, usage and storage of consumables, and parts
- management of expiry dates
- calibration activities
- metrological traceability
- quality assurance activities such as quality control, proficiency testing, etc.
- review and reporting of examination results;
- communication with clinicians

- communication around the intended use, contraindication, recalls, surveillance
- local network and internet connectivity;
- biohazardous waste disposal;
- equipment maintenance and servicing
- equipment decommissioning
- equipment disposal
- contingency and recovery planning

3. **Events under Control of the Clinician:** The events under control of the clinician normally begin with receipt and review of the examination result. This is subsequently followed by a decision as to whether to accept the result as valid or not. Large errors are likely to be questioned and rejected – this could include results contradicted by other diagnostic information or abnormal results which may prove to be incompatible with life. On the other hand, plausible results are likely to be accepted as valid and potentially used for medical decisions.

Point of care refers to the use at, for example, the physician's office, clinic or specialized hospital unit. When IVD medical devices are used at the point of care, clinicians are most commonly involved. Their involvement could be in the collection, handling, inventory, and storage of patient samples. These clinicians may perform many or all of the activities of the IVD medical device user. In these instances, the possibilities of the manufacturer influencing clinical activities through information for safety in accompanying documentation is quite great. The sequence of events identified for risk analysis should therefore reflect the use of the IVD medical device in the actual use environment.

Clinical uses contraindicated for, or not explicitly addressed in the accompanying documentation could be considered as reasonably foreseeable misuse for the purposes of risk management. It should be noted that accompanying documentation written for IVD medical device users in the laboratory does not always reach the clinicians who may only be ordering and acting upon the examination results.

Examples of decisions and activities typically under the clinician's control include:

- comparing the result to expected values
- requesting confirmatory or corroborating examinations;
- proceeding without the examination result (this could be if the results are delayed);

- establishing a clinical diagnosis;
- initiating or withholding therapy.

H.2.7. Estimation of the Probability of Occurrence of Harm:

H.2.7.1. General Considerations:

A risk analysis should consider the entire sequence of events of the device in question. This includes design and development, device labelling, manufacturing and supplier management, product inventory and distribution, equipment servicing and training and product support.

The probability of a patient being harmed is the combined probability of:

- each event in the sequence of events associated with a particular hazard; and
- the potential harm.

The approach used to estimate the probability of occurrence of harm can depend on the complexity of the sequence of events. For this reason, it is important that the analysis be performed by individuals with the appropriate knowledge and expertise in relation to the IVD medical device being analyzed. This should also include its analytical and clinical use as well as any technologies involved. For IVD medical devices that include a short sequence of events, the risk analysis can be relatively straightforward. A cross-functional team of experts can be assembled to develop estimates of the probability of occurrence of harm for each identified hazardous situation. This cross functional team acts based on their expert judgment, supplemented with available analytical and clinical information. This information could be from premarket studies, experience with similar products and/or relevant post-production information.

A cross functional team should generally include individuals with design, construction, use and servicing knowledge as well as knowledge of the use environment and an ability to interpret or understand the clinical use of the examination results.

In instances where software-containing devices are being assessed, the team should also include expertise in connectivity and security.

In instances where a sequence or combination of events are more complicated, segmenting the analysis at the hazardous situation level can result in more efficient use of expert resources. This allows for the application of their expert knowledge and expertise to the analysis of relevant sequences of events. This approach is called the "P1 x P2" approach.

A detailed explanation of the "P1 x P2" approach is provided below.

H.2.7.2. Guidance for Using the "*P*1 x *P*2" Approach as Provided by ISO 24971:

In instances where a sequence of combination of events could become more complicated, for example, where events extend beyond the medical laboratory; the application of the cross-function team's expert knowledge and expertise could be used in a P1 x P2 approach.

An illustration of this approach is presented below.

This approach is useful for complex risk scenarios with an extended sequence of events.

An example is given below:

Events from an incorrect result can extend beyond the medical laboratory to the decisions and actions of a clinician. This is often beyond any reasonable means of *risk controlled* by the laboratory or the *manufacturer*.

The probability that a *hazardous situation* would occur is given as *P*1 and the probability that *harm* would result from that *hazardous situation* is given as *P*2. These are estimated separately by appropriate experts. The probability that a hazardous situation (*P*1) would occur is related to the analysis in the laboratory using the *IVD medical device* and producing the result and the probability that harm would result from that hazardous situation (*P*2) is related to the use of the result by the clinician and the decisions and actions based on that result.

These probabilities are combined to obtain the overall probability of occurrence of *harm:*

$P = P1 \times P2$

For an *IVD medical device* intended for medical laboratory use, the sequence of events can be divided into analytical and clinical segments. The *hazardous situation can be* defined as:

- an incorrect result reported to a clinician
- a clinically significant delay in reporting the result, or
- failure to report an important examination result.

The probability of each segment can be estimated separately as follows:

- *P*1 is the probability that the *hazardous situation* would occur; and
- *P*2 is the probability that a specific *harm* would result from that *hazardous situation*.

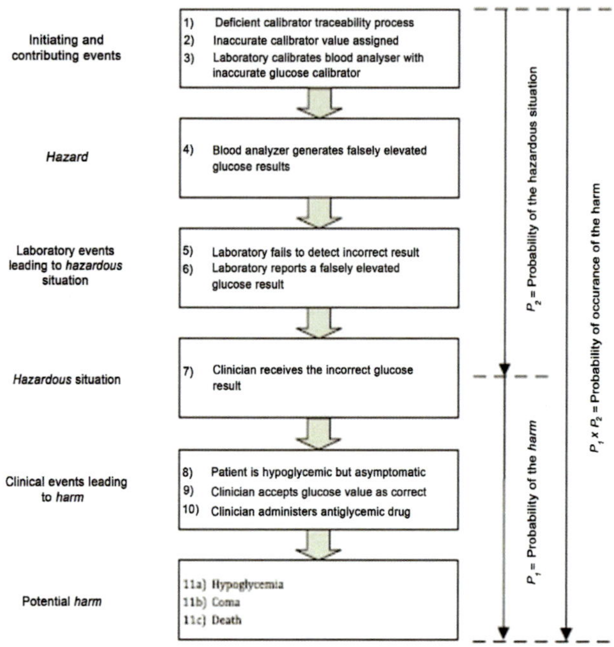

The above figure demonstrates one such way to apply the "P1 x P2" approach to an IVD medical device related risk scenario. The example used is a blood analyzer performing glucose measurements in a medical laboratory. The above figure depicts the entire sequence of events that could occur. This starts with:

- Failure of the manufacturer's calibrator value assignment process and ends with the possibility of multiple patient harms

1. **Hazard:** The hazard here is an incorrect glucose measurement result which shows as being falsely high and caused by inaccurate calibrator values which have been assigned by the manufacturer.

The first two events in this scenario are under the control of the manufacturer.

2. **Hazardous Situation:** The subsequent events leading to a hazardous situation occur in the laboratory and are beyond the manufacturer's direct control, but these are potentially controlled by information for safety provided by the manufacturer in the accompanying documentation.

The remaining events occur beyond the direct control of the laboratory. Therefore, the hazardous situation in this scenario can be defined as the event beyond any reasonable means of risk control by the manufacturer. For an efficient risk analysis in such cases involving incorrect IVD results, the hazardous situation can be defined as the event when the laboratory reports and/ or the clinician receives the incorrect result.

The probability of the hazardous situation occurring (P1) and the probability of the hazardous situation leading to harm (P2) can be estimated separately by the appropriate subject matter experts. The two probabilities can then be combined to give an estimate of the overall probability of harm. Let's look at this in more detail:

- **P1: Probability of Occurrence of the Hazardous Situation:** The individuals assigned to estimate the probability P1 should be familiar with design, construction, use and servicing of the IVD medical device. They should also have an adequate understanding of the use environment. The use environment includes, for example: the medical laboratory, point of care and/or patient's home.
- **P2: Probability of Harm Occurring from a Hazardous Situation:** The individuals assigned to estimate the probability P2 should be familiar with the medical use of the IVD results. Probability P2 can be estimated in the following ways:
    - through expert clinical judgment and experience with similar IVD examinations;
    - informed by adverse event data;
    - medical literature; and
    - information from post-production.

Detailed understanding of the performance of the IVD medical device or how the results were generated and reported is generally not needed to estimate P2.

H.2.7.3. Guidance for Estimating the Probability of Occurrence of Harm as Defined by ISO 24971:

The below table is intended to guide the development of suitable probability estimates. The table presents a series of questions which should be adapted as appropriate for the type of IVD medical device, the specific intended use and the risk estimation approach used.

Questions 1 to 4 pertain to the analytical portion of the sequence of events. These questions can help a manufacturer estimate P1.

Questions 5 to 8 pertain to the clinical portion of the sequence of events. These questions can help a manufacturer estimate P2.

4.28.2. Questions to Help Estimate the Probability of Occurrence of Harm

What is the Likelihood That:	Points to Consider
• The initiating event would occur (i.e., a device failure or use error)?	• How effective are prevention/detection measures? • Can probability be estimated? (If not, set probability = 100%) • Would frequency depend on use environment? (Address worst case) • Can specific faults, failure modes and/or use errors occur in a reasonably foreseeable combination to cause a hazard?
• An incorrect result would be generated by the IVD medical device failure or use error?	• How effective are measures intended to ensure accurate results? • Or detect an unacceptable change in analytical performance? • Would conventional quality control procedures cause the incorrect examination results to be rejected? • What is the influence of the use environment (Different use environments should be analyzed separately) • Would the device prompt a user to correct the problem in time to obtain a valid examination result upon repeat?
• The incorrect result or incorrect ancillary information would be reported to the clinician?	• Are abnormal results for the examination reviewed against critical value limits • Are abnormal results verified prior to reporting to the clinicians? • Are rare or unexpected results automatically confirmed prior to reporting?

• A clinically significant delay in reporting the examination result would occur? • A clinically significant delay in ancillary patient information would occur?	• Is the result critical for a timely diagnosis? • Is the result critical for a therapeutic decision? • How much of a delay would create a hazardous situation? • Would the time necessary to troubleshoot a malfunction or out of control situation cause a clinically unacceptable delay? • Would a backup procedure to ensure timely availability of results be an expectation of standard laboratory/medical practice? • Could a second examination be performed and the result be reported within the time required for a critical result?
• The clinician will believe the incorrect result to be valid?	• Would a clinician recognize the result as incorrect for reasons such as 1) inconsistency with a patient's clinical status, 2) contradicted by other clinical data, or 3) being physiologically implausible? • Would a competent clinician question, repeat or corroborate a result that did not fit the clinical impression? • Do current standards of medical practice require confirmation?
• An incorrect medical decision and/or intervention (or lack of intervention) will occur due to the incorrect result?	• Are the results used for diagnosis, therapy or monitoring? • Will the result be the primary basis for a particular medical decision? • Will the result only be used in the context of signs, symptoms, other examination results and the patient's medical history? • Do positive or abnormal results always lead to a particular medical decision or treatment? • Do positive or abnormal results always lead to further investigation? • Would a false negative or false "normal" screening result cause the clinician to miss a treatable medical condition?

• An inappropriate medical decision or action will be caused by failure to receive a timely IVD result?	• To what degree is the result used to guide the intervention or therapy? (This is based on the signs, symptoms, medical history and other examination results that would be available to the clinician).
• Patient harm will be caused by the inappropriate medical decision or action?	• How urgent is an immediate decision or intervention for the patient? • What are the medical consequences of the inappropriate action or delay in taking necessary action? • To what extent would the condition of the patient increase the probability of occurrence of harm? • Are there implications for individuals other than the patient. This may include: o potential for transmission of infectious agents to others? o exposure of an embryo or fetus to teratogenic agents or radiation? o antimicrobial resistance due to unnecessary exposure? o false rejection of an organ for transplant? o need for family counselling due to a false diagnosis? o parental anxiety from false positive newborn screening result?

4.28.3. Risk Control as per ISO 24971

It is important to remember that the decisions and actions of clinicians are out of any reasonable means of risk control by the manufacturer. For this reason, risk control activities should focus on reducing the probability of events under the control of the manufacturer.

This ideally should include:

- providing information for safety purposes
- verifying the effectiveness of information for safety to users in the laboratory

H.3.2. Inherently Safe Design and Manufacture as per ISO 24971:

Risks to patients are generally reduced by:

- lowering the probability that clinically significant delays will occur through ensuring performance characteristics meet medical requirements;
- by lowering the probability that incorrect results will be reported.

In such instances for quantitative measurements of analytes, limiting the magnitude of errors can in turn lower the frequency of inappropriate medical decisions. Such analytes include blood glucose, electrolytes, enzymes and therapeutic drug.

Design features that control accuracy and reliability of the examination results are listed below:

- trueness of the calibrator values
- traceability to a recognized reference standard
- measurement uncertainty through precision of the measuring system
- analytical specificity of IVD reagents
- detection limit or quantitation limit
- reliability of the instrument
- discrimination between positive and negative samples
- eliminating mistake-prone procedural steps through automation and mistake-proofing
- component version traceability and positive sample identification
- software functionality
- system ease of use
- data network and internet connections
- reduced reagent or calibrator variability
- reduced lot-to-lot specifications and supplier requirements
- prevention of intermittent component failures
- stability of reagents, calibrators or control materials
- microbiological control

H.3.3. Protective Measures in the IVD Medical Device or Manufacturing Process as per ISO 24971:

The ISO 24971 standard includes examples of detection features in the IVD medical device or reagent kit which is intended to prevent conditions that can cause incorrect or delayed results. These include:

- liquid level sensors to ensure sufficient sample volume are measured
- fault detection systems such as inadequate temperature control
- sample quality checks
- controls to detect and remove sample artefacts
- built-in controls to verify correct calibrator or reagent lots
- alarms and error messages to alert users to fault conditions and recovery procedures
- software that identifies questionable results for reflex testing, review or suppression
- incoming inspections of supplied components
- in-process acceptance testing and final-product acceptance testing

H.3.4. Information for Safety as per ISO 24971:

Information for safety purposes is provided to users of IVD medical devices to prevent the occurrence of a hazard or a hazardous situation. This can be through an effective risk control measure if the following occurs:

- information instructs users on what actions to take or avoid;
- the intended users are capable of following the instructions;
- it can be reasonably expected that they will follow those instructions.

In addition, consequences of ignoring the information for safety should also be clear. This information for safety should be used in the risk management process. Examples of information for safety that enable users to control risks include warnings, instructions and other information. This information should address:

- chemical or biological hazards associated with the IVD medical device
- contraindicated medical conditions or clinical applications
- sample collection, storage and preparation
- identification of inappropriate sample types
- interfering substances detectable by the user

- causes of hazards, including potential use errors
- incompatible system components and accessories
- utilities and facilities where the IVD medical device are to be installed (e.g., use environment)
- improper reagent storage or use beyond the expiry date
- installation, servicing and disposal of the IVD medical device
- quality control samples and frequency
- validated measuring intervals and dilution instructions for samples when the measured values are above the upper limit of the measuring interval
- biological reference intervals and medical decision points
- validated cleaning methods for reusable items
- preventive maintenance procedures
- interface and connectivity requirements
- backup and recovery in case of system failure

H.3.5. Role of Standards and Analytical Performance Criteria as per ISO 24971:

There are a few international product standards which define the generally acknowledged state of the art for IVD medical devices, for example, ISO 15197 and ISO 17593.

ISO 15197 is the standard for self-monitoring devices for blood glucose and ISO 17593 is the standard for self-monitoring devices for oral anticoagulant therapy.

In addition, there are some international standards for IVD medical devices which address certain aspects of inherent safety which can provide evidence that the risks from specific hazardous situations have been reduced to the state of the art. These include ISO 17511 for example. ISO 17511 defines a process for establishing the metrological traceability of IVD calibrator values in relation to higher order reference materials which define the state of the art for accuracy of patient results. Other examples of IVD standards with potential relevance to risk control include:

- ISO 17822-1 (nucleic acid-based detection systems)
- ISO 20776 (antimicrobial susceptibility)
- ISO 20916 (clinical performance studies)
- ISO 23640 (stability of IVD reagents)

- IEC 61010-2-101 (safety of IVD equipment)
- IEC 61326-2-6 (electromagnetic compatibility of IVD equipment)
- IEC 80001-1 (networked medical devices)

It is also important to note that the manufacturer is also responsible for justifying the extent that such standards and performance criteria apply to their particular IVD medical device and its intended use as well as for verifying that all relevant hazardous situations have been considered.

H.3.6. User Education and Training as per ISO 24971:

In instances where user interfaces are quite complex, examination procedures are difficult or critical clinical applications are involved, information for safety can take the form of training and education programs. These programs can be offered by the manufacturer to help avoid use errors. Training materials suitable for use in continuing education programs can also be provided.

In determining the degree of risk reduction related to information for safety, consider the following:

- the use environment, competence and capabilities of device users can vary widely
- quality control and quality assurance practices are not uniform around the world
- information about contraindicated medical use and interfering drugs provided to IVD medical device users might not always reach the clinicians who order the examinations.

H.4. Benefit-Risk Analysis as per ISO 24971:

In instances where it is not possible to describe the benefits of an IVD medical device directly, surrogate endpoints can be established. Examples of surrogate endpoints include the ability of an IVD medical device:

- to identify a specific disease
- to provide diagnosis at different stages of a disease
- to predict future disease onset and/or
- to identify patients likely to respond to a given therapy

H.5. Disclosure of the Residual Risks as per ISO 24971:

The aim of disclosing residual risks is to provide information to the device user and potentially the clinician and the patient such that they can weigh the risks of using the IVD medical device against its benefits and make informed decisions about the risk acceptability. It is also important for manufacturers to take into account the information needed by medical laboratories and clinicians so as to evaluate the inherent residual risks and determine the need for further risk reduction measures.

Disclosure of residual risks for IVD medical devices can occur in various different ways. This includes:

- information provided in the accompanying documentation about the performance specifications and claims
- limitations of the IVD medical device or examination procedure
- potential causes of hazards and hazardous situations that could not be eliminated by the manufacturer

The disclosure of residual risks is in addition to the risk control measures provided to users as information for safety.

H.5.2. Performance Specifications as per ISO 24971:

Description of the relevant analytical performance characteristics and specifications s well as the results of clinical performance studies allow for an evaluation of the utility of the IVD medical device for its intended medical applications.

The description of the performance characteristics should allow laboratory or other users to:

- verify that the IVD medical device is performing as intended by the manufacturer
- determine the measurement uncertainty associated with the examination results
- know that the examination results will meet the medical needs of the clinicians

H.5.3. Limitations of the IVD Medical Device as per ISO 24971

The ISO 18113-1 requires that any analytical and clinical limitations be disclosed. The limitations describe situations in which the IVD medical device might not perform as intended and as such, may disclose residual risks. These include:

- interfering substances which are not detectable by the user such as drugs, biological metabolites, etc.
- specific patient populations in which the performance characteristics might not apply

- values outside the measuring interval where performance characteristics are not validated
- patient populations where reference intervals or medical decision points might not apply
- primary sample types that have not been validated for the intended use
- circumstances and factors that might affect examination results, but have not been studied

H.5.4. Generally Recognized Limitations of Use as per ISO 24971:

Events or circumstances which could lead to a hazardous situation may also present some risks. These are addressed by standard medical or laboratory practices and are typically not described in the instructions for use. This done to avoid overwhelming users with unnecessary details.

In such instances, manufacturers should consider if additional information is needed to ensure that users are aware of risks associated with such events or circumstances. They should also keep in mind the hazard associated with communicating such risks. In instances, these users might have difficulties understanding which ones are important to control.

H.6. Production and Post-Production Activities as per ISO 24971:

It can be fairly difficult for manufacturers to establish an effective risk management system that monitors post-production information. This information includes complaints, adverse events and product nonconformities. This difficulty arises, more so for devices intended for use in medical laboratories. This is because reliable data to monitor the actual frequency of hazardous situations and harms can be difficult to obtain.

An easier route is monitoring the occurrence of hazards and their causes as this feedback can be provided directly by the device users who experience the events.

It is important for reports of device failures, use errors and medical incidents to be collected and analyzed. In addition, the observed frequencies should be compared to the anticipated frequencies.

A system should be established for collecting post-production information. In this instance, manufacturers can use the product risk analysis to develop a classification and coding scheme for anticipated harms, hazardous situations, hazards, and their causes.

Events can be classified according to the estimated severity of the potential harms, as well as the probability that incorrect or delayed results would lead to harm. This type of a system can be useful for complaint handling, post-market surveillance and adverse event reporting as well as product failure investigations.

H.6.2. Monitoring Analytical and Clinical Performance as per ISO 24971:

For a system to be effective in collecting production and post-production information for IVD medical devices, there should be monitoring of analytical performance data from both internal and external sources. Internal sources of performance data include:

- production data which may be routinely collected during product release testing
- value assignment activities
- stability monitoring processes
- product failure investigations

External sources of performance data on the other hand can include:

- data routinely obtained from voluntary participation in external quality control and proficiency testing programs
- user performance evaluations
- instrument installation
- servicing activities

H.6.3. Monitoring Clinical Performance as per ISO 24971:

Medical laboratories generally have no way to know that a reported result may be incorrect and which could in turn lead to an inappropriate medical decision, intervention or injury. This is only possible if they receive a complaint from a clinician. For this reason, the severity of any reported harms as well as their frequency should be carefully monitored as part of post-production activities.

In addition to monitoring customer feedback about clinical incidents, the manufacturer should also attempt to learn about new and emerging hazards or hazardous situations occurring with similar products. These sources include:

- adverse event and recall databases maintained by regulatory authorities
- reports from public safety institutes
- national medical laboratory associations
- medical literature

CHAPTER 5

BUILDING A RISK MANAGEMENT FILE

Contents

5.1. WHAT SHOULD GO INTO YOUR RISK MANAGEMENT FILE?

A risk management file is where all risk management activities, documentation, and records are maintained. Essentially, a Risk Management File should contain the following documentation:

- Risk management plan
- Risk analysis
- Risk evaluation
- Risk controls
- Evaluation of overall risk acceptability
- Risk management review
- Production and post-production risks

The Risk Management File can be built for an individual product or for a product family. This decision will depend on the business building the Risk Management File and how their product line is positioned.

5.2. BUILDING A RISK MANAGEMENT FILE

Building a Risk Management File is similar to building a Design History File or a Technical File. The Risk Management File should contain all the evidence that shows hazards are identified, mitigated, and further evaluated once mitigations have been implemented. In addition, traceability should be maintained for each hazard identified to the associated risk analysis, risk evaluation, risk controls, and evaluation of residual risks. A simple example is demonstrated below:

Hazard	Risk Analysis/Risk Evaluation	Risk Controls	Implementation and Verification of Controls	Evaluation of Residual Risks
Bacterial contamination	Risk assessment RA-0001	Sterilization	Validation protocol 0002 for sterilization	RA-0002, residual risk analysis

The above table shows a sample hazard of bacterial contamination. An individual could then consult each referenced document to verify that the risk was adequately controlled and that any residual risk was found to be acceptable or, alternatively, document that the risk-benefit rationale demonstrates that the benefit outweighs the risk.

5.3. RISK MANAGEMENT PLAN

A risk management plan is a document associated with the device in question. It should identify various risk management activities anticipated and planned throughout the product's life cycle. A Risk Management Plan should include the following criteria:

1. Scope of the Risk Management activities – this should include a definition of the products included. It is possible to have multiple products described within a single Risk Management Plan.
2. The intended use of the product or product family.
3. Identification of all Risk Management activities planned throughout the product life cycle.
4. Definition of the roles and responsibilities. This section should include identification of the Risk Management team that will be reviewing and approving risk documentation.
5. Criteria for the product's risk acceptability.
6. Methods to verify Risk Control measures and reduce risks to the pre-established acceptable levels.
7. Definition of how post-production information will be captured and fed into Risk Management activities for the product.

It should also be noted that the Risk Management Plan evolves over time. Despite this, it should be kept up to date and current, even after product development is completed.

The Risk Management Plan should be one of the first documents in the design and development process. Each medical device must have its own Risk Management Plan, which should identify the risk management activities that need to occur at each phase of the product life cycle.

In addition, the Risk Management Plan needs to establish how the device risks will be evaluated to determine if the risks are considered acceptable. This should result in the plan being updated at regular intervals. The plan should ideally serve as a guide for the who, what, where, and when of the risk management process for the specific device. Activities that need to occur should be clarified in each phase of the life cycle, from design to post-market. It is important to remember that the plan is a living document that should be periodically reviewed. The initial plan should clearly lay out who needs to conduct each of the risk management activities and when it must be done. Timing is especially important for risk activities that build upon one another.

The next part of the Risk Management Plan should focus on how the process will be applied. It should include the following pieces of information:

- A review of the requirements for each risk management activity
- Risk acceptability criteria
- Residual risk acceptance criteria
- Plan for verification of risk controls
- Plan for collecting and reviewing post-production data
- A review of the requirements for each risk management activity

To help complete the above required information, below are some questions to prompt the required information:

• Risk acceptability criteria	• What levels of risk are acceptable? • How will acceptability be determined if a quantitative probability cannot be assigned to the hazard? • Risk acceptability criteria should be defined before risk analysis and estimation to keep the process as objective as possible.
• Residual risk acceptance criteria	• What level of residual risk is acceptable? • How will residual risk be identified and determined?
• Plan for verification of risk controls	• How will it be verified that risk controls have been implemented into the process? • How will these controls be reviewed to determine effectiveness at reducing risk?
• Plan for Collecting and Reviewing Post-Production Data	• What sources will be used to gather post-production data? • How will that data be reviewed? • How will it be used as an input to the ongoing risk management process?

5.4. RISK ACCEPTABILITY CRITERIA

It is important for manufacturers to define the risk acceptability criteria right at the start of the risk management process. This will aid in making the risk management process as objective as possible. If the risk acceptability criteria are established early in the process, such as in the design phase, the criteria are less likely to be influenced by data that's been acquired during the development process, which can cause the process to become subjective. Criteria can then be quantitative values based on a calculation of a risk index number. This risk index number should be calculated from the probability, severity, and any other metrics used to quantify the risk of potential harm. The risk acceptability criteria should be applied before any mitigations or risk controls are put into place, so a higher level of risk is generally acceptable at this point.

5.5. RESIDUAL RISK ACCEPTANCE CRITERIA

The residual risk acceptance criteria are similar to the risk acceptability criteria; however, the only difference is when the criteria are applied. The residual risk acceptance criteria will only be applied after risk controls have been implemented and will be applied only to residual risks estimated after all mitigations have been completed.

5.6. PLAN FOR VERIFICATION OF RISK CONTROLS

The plan for verification of risk controls will identify how mitigations will be verified. This process should be similar to verifications completed if a design change is implemented through the change control process. A verification plan should be in place, and this should consider whether any validation activities need to be completed. This should be done to ensure that the mitigation did not cause any unintended issues with another aspect of the process or design. The verification plan should also aid in determining if the mitigation was effective in reducing the probability of harm from occurring.

5.7. PLAN FOR COLLECTING AND REVIEWING POST-PRODUCTION DATA

This is a recent requirement that many manufacturers are not quite familiar with yet. Traditional processes for receiving feedback for the risk management process are generally through the process and product nonconformance system and the customer complaints system. This channel, however, takes it one step further by including the analysis of information from all levels of the supply chain, as well as the analysis of information about the current state of the art and any publicly available information. Post-production information would generally include literature reviews, regulatory database searches for recalls, and so on.

5.8. RISK ASSESSMENT

There is a simple equation that should be remembered when risk assessment is concerned:

Risk assessment = Risk Analysis + Risk Evaluation

The starting point for identifying specific risks related to a medical device or related processes is the risk analysis. This could be done in a number of ways and through various techniques, such as preliminary hazards analysis, FMEA, and fault tree analysis. Each of these techniques has its pros and cons, and the best method for a business should be determined and chosen appropriately. The Risk Analysis Process should be typically described in the company's risk management procedure. It should identify the medical device, as well as who was involved, the risk analysis scope, and related dates. Having a defined intended use of the medical device is imperative, as this will ensure the appropriate controls are put in place. A manufacturer should ensure that the intended use is the same as that used when initial design controls, user needs, and design inputs were established. Knowing the intended use is important for Risk Management as this statement helps define the scope and will be instrumental in identifying the related hazards, harms, and so on. Once the intended use is defined, cases of foreseeable misuse will also be able to be identified. In addition, off-label uses of the device should also be considered. Safety characteristics should also be considered.

Identification of hazards is also important, as these are potential sources of harm. A manufacturer will need to identify all the possible hazards, and the ISO 14971 standard (Annex C) is a great way to start, as it contains a good list of examples of hazards. Common examples include:

- Electromagnetic energy
- Line voltage
- Leakage current
- Electric fields
- Thermal energy
- Mechanical energy
- Gravity
- Vibration
- Biological

- Bacteria
- Viruses
- Chemical

Once again, understanding the intended use is important when it comes to identifying hazards and foreseeable misuse. One such example is to go through all the steps required for the device to be used. At each step, the manufacturer should identify if there are any potential sources of harm. This harm can be viewed from the perspective of the patient, as well as end-users, damage to property, and the environment.

5.9. RISK ANALYSIS

Each medical device should have its own unique risk analysis. A risk analysis for a similar device or device family can also be used as a starting point for the analysis, but this should not preclude further device risk analysis. The analysis should be done by a cross-functional team to ensure that a multi-disciplinary approach is adopted. This could be in the form of a brainstorming session or working collaboratively on a shared risk management file. Essential elements of the risk analysis should include an identification and description of the device being analyzed, who was involved in the analysis, and the scope of the analysis. In addition, team members with greater knowledge of certain areas should be designated to analyze risks in those areas in relation to the intended use and foreseeable misuse. For example, the quality/regulatory team may be better suited to look at safety since they are familiar with complaints, recalls, and the relevant standards. An initial list of hazards should be constructed based on the intended use, foreseeable misuse, and safety characteristics. Thereafter, the team should work collaboratively to identify hazardous situations.

5.10. SOURCES FOR RISK ANALYSIS

There are various sources that could be used to determine hazards and hazardous situations. These include publicly available information sources about similar devices already on the market, complaint reporting data, and so forth.

5.11. RISK ESTIMATION

Risk estimation can be identified in the risk assessment either qualitatively or quantitatively, or possibly through a mix, depending on the availability of quantitative data. For each hazardous situation that is identified, a risk estimation must be applied. Estimating the probability of occurrence of harm is not possible for all hazardous conditions. Therefore, in some instances, it is possible to simply identify the possible consequences. The risk estimation needs to consider both the probability of harm occurring and the severity of the harm. From here, a risk index level should be assigned to that hazardous situation. This could be calculated using an equation, or for a more qualitative system, it could be assigned a general category of risk level. To estimate risks, even qualitatively, there should be some sort of data or information to support that assessment. These sources may vary widely depending on the hazard under evaluation and how common a hazardous situation is in the industry. Data from publicly available incident reports or published literature are easy to obtain and can be very specific, especially if there is already an existing device that is similar.

5.12. RISK EVALUATION

For the risk evaluation to be completed, it is imperative that the acceptability criteria are well defined. If the criteria are clearly defined, it should be a quick task to compare the estimated risk level to the criteria to determine if it passes or fails. If the risk fails to meet the acceptability criteria, the manufacturer should then attempt to put adequate risk controls in place. The risk evaluation step should be revisited once risk controls are implemented and verified.

5.13. RISK CONTROL

When a risk is identified, especially if the hazardous situation is likely to occur, risk control measures should be taken to reduce the risk of the hazardous situation actually occurring. The type of control applied can vary widely based on the hazardous situation. These controls could include training, labelling, validation, and even design process characteristics. The first step in applying risk control would be to ensure appropriate materials are selected when designing the device. For example, non-toxic or biocompatible materials. The next step should ensure that features of the device or in the manufacturing process that are protective, such as safety guards, product markings, or quality control checks, are in place. The final step of control should involve information for safety, including labelling, instructions for use, and user training. This last step of control is the least effective because it relies on the user to do the right thing, rather than the manufacturer implementing preventive measures.

5.14. IMPLEMENTING RISK CONTROLS

Once you have determined which risk control measures need to be implemented to reduce risk, they should be implemented and then further verified. This could be done through process qualifications and validations, for example, which could form part of the standard design and development process. For mitigations that aren't already being verified through another process, the manufacturer will need to deliberately document that the measure has been implemented. Furthermore, the verification of that implementation should include verification of the effectiveness of the control measure if possible. The evidence of the implementation and verification activities must form part of the Risk Management File and be traceable to each hazardous situation identified during your risk analysis.

5.15. EVALUATING THE RESIDUAL RISK

The last step in the risk process is to evaluate the residual risk and ensure that a hazardous situation is reviewed after risk control measures have been implemented and verified. This evaluation process considers the risk that still remains, not only for the individual hazards and hazardous situations but for the overall device. The residual risk should take into account the expected benefits that the device provides to the user or patient when used in accordance with its intended use. Keeping these benefits in mind, the residual risk should be compared against the acceptability criteria to determine if the overall device risk is acceptable. This should then be related back to the predefined acceptability criteria.

The overall residual risk may be documented by identifying the residual risk for each hazardous situation, then adding a statement about the overall risk-benefit rationale and identifying whether the risk was found to be acceptable. Any significant risks that still remain must be identified and disclosed. These residual risks should be mentioned in the Instructions for Use (IFU) for the device.

5.16. RISK MANAGEMENT COMPLIANCE REVIEW

The last step of the risk management process, which occurs prior to device release, should be a comprehensive review of the entire process. This review should be built into the design controls system as one of the final steps before pushing the device to the clinical stage or commercial production stage. This process will most likely be assigned to the Quality or Regulatory team and will be much like an audit of the process. This final risk review will verify that the risk management plan has been established and implemented, including the steps for risk management that will occur once the device goes into production. In addition, there should be a review of systems to ensure that procedures are in place for post-production risk management activities. This review will also verify that the mechanisms for collecting and reviewing post-production data have been implemented.

Finally, the risk review will also verify that the residual risk evaluation is documented and that the findings concluded that the residual risk is acceptable. This will be confirmed by reviewing the documentation, including the Risk Management File, and compiling a risk management report and/or a risk-benefit report.

CHAPTER 6

HOW TO CONDUCT AN ONGOING RISK MANAGEMENT PROCESS?

Contents

6.1. DATA COLLECTION

Conducting a risk assessment is an ongoing process that should run from design conception through to post-production. Manufacturers should establish channels for collecting post-production data within their risk management system. This can start with the Risk Management Plan. Since the Risk Management Plan is a living document, it should be periodically updated as new sources of information are established. The following are some examples of information sources:

- Information from the production process:
  - Nonconformance data;
  - Receiving inspection data;
  - Data from any production tests that are routinely run.
- Information from the device user:
  - User information;
  - Customer surveys;
  - Device packaging.
- Information from installation and maintenance personnel:
  - Issues encountered during installation and maintenance;
  - Personnel maintaining sensitive electronic equipment;
  - Hospital personnel cleaning and sterilizing devices for reuse.
- Information generated by the supply chain:
  - Distributors can pass on information from their customers.
- Publicly available scientific literature and regulatory reporting:
  - Device incident data;
  - Scientific literature.
- Information about the state of the art:
- This type of information comes from conferences, technical journals, and industry publications;
- Newer competing devices.

6.2. DATA REVIEW

Collecting data and establishing data collection channels is the most difficult part of the post-production risk management process. Once a manufacturer has the data and information, it can be fed back into the risk management process, starting again at the risk analysis. To do this, a manufacturer will need to establish a process for periodic review of the data and information along with a review of the existing risk analysis documents. The review process should also include escalation points for immediate evaluation of the risk analysis documents if concerning data or information is received. For routine review of the post-production data, the risk analysis team will need to sift through the new information and data to determine if there are new hazards or hazardous situations that were not previously captured in the risk analysis. Additionally, they will need to determine if new information suggests that the documented risk estimate may not be accurate.

6.3. REVIEW OF OUTPUTS

When risk management input data is reviewed, it should be documented and added to the Risk Management File. This documentation should include any decisions or actions that were taken. This can be done through meeting minutes, risk reviews, or any other documented form. The documentation should also include any revised copies of the risk assessment documents, such as the risk analysis, risk estimation, risk controls, and residual risk evaluation. The Risk Management Plan may also need to be updated if changes in the state of the art indicate that the risk acceptability criteria should be changed or if other changes to the plan are needed.

The output from this review process should then be fed into the management review process as an input. This information can be used by management to determine if the process is suitable and effective.

CHAPTER 7

ANNEXURES

Contents

7.1. SIMPLIFIED RISK POLICY TEMPLATE

- Risk Policy of Business A:

The management of risk will be undertaken through a systematic process focusing on the guidance of the International Standard for Risk Management: ISO 14971:2019 relating to medical devices as a function of the ISO 13485:2016 Quality Management System.

1. **Purpose of the Risk Policy:** This aim of this risk policy is to ensure the safe design, development, manufacture, and performance of Business A's medical devices by defining a suitable process for management of risk as well as criteria for risk acceptability which are fully integrated into the businesses Quality Management System.
2. **Introduction to the Risk Policy:** The ISO 14971:2019 standard requires that top management provide evidence of their commitment to the risk management process through provision of adequate resources, assignment of competent personnel for risk management and through establishment of criteria for risk acceptability.

The policy shall ensure that the criteria established are based upon the relevant applicable national or regional regulations as well as relevant international standards and regulations. In addition, the risk management system shall take into account available information such as the generally accepted state of the art and known stakeholder concerns.

The criteria for risk acceptability must include criteria for accepting risks when the probability of occurrence of harm cannot be estimated. In establishing the policy for determining risk acceptability criteria, the ISO 14971:2019 standard allows manufacturers to choose from a range of options within those permitted by regulations.

European Medical Devices Regulations require that, in selecting the most appropriate solutions for the design and construction of the devices, these solutions must conform to safety principles, taking into account the generally acknowledged state of the art, and the manufacturer must apply the following principles in the following order:

- Eliminate or reduce risks as far as possible (inherently safe design and construction),
- Where appropriate take adequate protection measures in relation to risks that cannot be eliminated,
- Inform users of the residual risks due to any shortcomings of the protection measures adopted.

In this context, 'eliminating' or 'reducing' risk must be interpreted and applied in such a way as to take account of technology and practice existing at the time of design and of technical and economic considerations compatible with a high level of protection of health and safety.

3. **Objectives:** The objectives of the Risk Management Policy are to:

- Promotes and support risk management practices across the organization through implementation of a framework that provides a process to identify, analyze, assess and prioritise all areas of risk based on the following core risk areas:
  - o Organizational risk management
  - o Product and service delivery
  - o People
  - o Financial and economic
  - o Technology and information management
  - o Ethics
- Have a risk management plan for company products that will include product risk and post market surveillance.
- Provide adequate training, information and supervision to staff
- Develop and maintain the QMS in line with Quality Standards and Directives.
- Ensure efficient resources and operational capabilities are identified and deployed effectively.

4. **Risk Acceptability Policy:** Business A shall make every effort to ensure that its medical devices meet their indications for use and intended use. The safety and efficacy of the devices shall be well established. Each device shall have a suitable medical device file inclusive of a Clinical Evaluation whereby the safety and performance objectives and related specifications and measurables are detailed. The clinical evaluation shall serve as a base for addressing the potential risks regarding user safety and efficacy.

Each risk shall be addressed and its acceptability determined when weighed against the intended benefits of the device. Risk analyzes shall include technical and clinical aspects relating to the particular device concerned. The intended benefits of the device shall also be related to the accepted benefits associated with the generally acknowledged "state of the art".

ANNEXURE B

7.2. RISK PLAN TEMPLATE 1

Risk Plan of Business A

- **Purpose:** The purpose of this document is to ensure that the risk management activities carried out by Business A are planned and aligned with the relevant regulations and standards as defined below.
- **Scope:** The risk plan is applicable to risk assessments conducted on medical devices for Business A throughout all life cycle phases of the product. This includes planning, design and development, manufacture, packaging and labelling, sales and marketing, distribution, post market and disposal.
- **Definitions:**
    - **Harm** – Injury or damage to the health of people, or damage to property or the environment
    - **Hazard** – Potential source of harm
    - **Life Cycle** – Series of all phases in the life of a medical device from the initial conception to final decommissioning and disposal
    - **Residual Risk** – Risk remaining after risk control measures have been implemented
    - **Risk** – Combination of the probability of occurrence of harm and the severity of that harm

Abbreviations:

None

Roles and Responsibilities: The Risk Assessment Team of Business A shall consist of the following persons:

- The Managing Director;
- General Manager;
- Quality Manager;
- Regulatory Manager
- Group Engineer;
- Other personnel from various departments (sales, production, etc.) as required.
- External consultants (where necessary)

Business A shall ensure that the members of the Risk Assessment Team shall have sufficient knowledge and experience in risk assessment and management techniques and application of the device. Suitable records of competence shall be maintained.

Applicable Documents:

(Business A to list all applicable standards and regulations used when writing the risk plan here)

Overview of Procedure for Risk Management:

The risk management process shall be managed as follows:

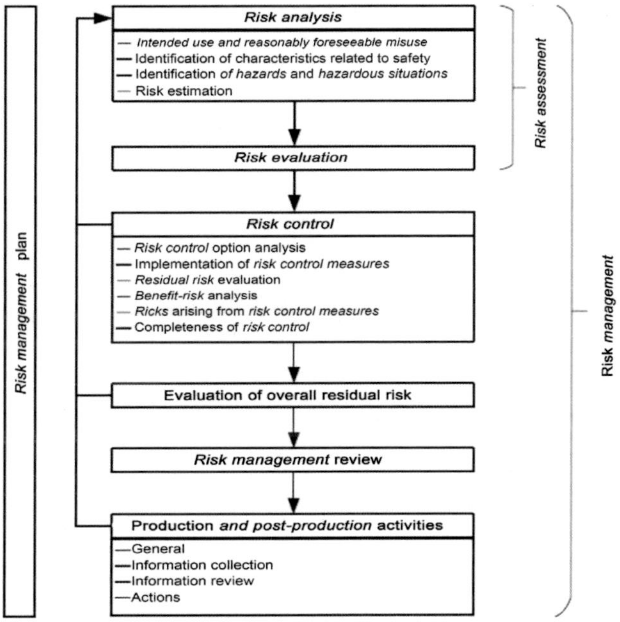

Requirements for Review of Risk Management Activities: A risk review shall be performed annually or as needed if previously unrecognized hazards or hazardous situations are identified.

A review and update of the General Safety and Performance Requirements Checklist should be conducted in the event of a review of the risk management activities.

The risk management file shall include:

- Risk policy
- Risk plan
- Risk assessment matrix

- Risk management report
- Risk benefit report

The Management File shall be updated where indications of new or increased risks are obtained. These risks may arise from:

- Internal non-conformances;
- Customer Complaints;
- Inputs from Product Development activities
- Input from the post-market surveillance system
- Inputs from the vigilance system (inclusive of recalls, field safety corrective actions, adverse events)
- Where major changes are done to the product design or product specifications
- Where the class of the medical device is changed
- Where raw materials are changed
- Where packaging materials are changed;
- Where the sterilization process is changed and
- Where raw material suppliers or subcontractors are changed

Criteria for Risk Acceptability: Criteria for risk acceptability shall be determined by applicable regulatory requirements, relevant international standards, generally acknowledged state-of-the-art and any validated concerns from appropriately identified stakeholders. Estimation/quantification of risks for each hazard shall be based on the worst-case scenario/single fault condition.

Severity rating table: (risk acceptability criteria)

Severity	**Rating**	**Description**
Life threatening	10	Death has or could occur
Severe	8	Serious injury, permanent impairment, irreversible injury
Moderate	6	Significant impairment that is either temporary or reversible
Limited	4	Transient/self limiting illness or injury
None	2	No injury or adverse health consequence anticipated

Probability of Occurrence	Rating	Qualitative Estimate Description	Quantitative Estimate Probability
Continuously Occurring	10	Occurrence is almost inevitable.	≥10% As frequent or more frequent than 1 in 10
Frequent	8	Occurrence is expected to occur repeatedly.	≥1%, but less than 10% As frequent or more frequent than 1 in 100 but Less frequent than 1 in 10
Occasional	6	Occurrence is expected to occur occasionally.	≥ 0.1%, but less than 1% As frequent or more frequent than 1 in 1,000 but Less frequent than 1 in 100
Rare	4	Occurrence is expected to occur relatively infrequently.	≥ 0.01%, but less than 0.1% As frequent or more frequent than 1 in 10,000 but Less frequent than 1 in 1,000
Remote	2	Occurrence is not likely to occur.	less than 0.01% (<1:10,000) Less frequent than 1 in 10,000

Probability of Occurrence of Harm	Severity				
	2 **None** (No injury or adverse health consequence anticipated)	**4** **Limited** (Transient / self-limiting illness or injury)	**6** **Moderate** (Significant impairment that is either temporary or reversible)	**8** **Severe** (Serious injury, permanent impairment, irreversible injury)	**10** **Life-Threatening** (Death has or could occur)
10 **Continuously Occurring** (almost inevitable)	**As Low as Possible**	**As Low as Possible**	**Unacceptable**	**Unacceptable**	**Unacceptable**
8 **Frequent** (expected to occur repeatedly)	**Acceptable**	**As Low as Possible**	**As Low as Possible**	**Unacceptable**	**Unacceptable**
6 **Occasional** (expected to occur occasionally)	**Acceptable**	**Acceptable**	**As Low as Possible**	**As Low as Possible**	**Unacceptable**
4 **Rare** (expected to occur relatively infrequently)	**Acceptable**	**Acceptable**	**Acceptable**	**As Low as Possible**	**As Low as Possible**
2 **Remote** (not likely to occur)	**Acceptable**	**Acceptable**	**Acceptable**	**Acceptable**	**As Low as Possible**

All risks shall be assessed to determine methods to minimize or eliminate the risk. Where the residual risk cannot be reduced or eliminated and the benefits outweigh the risk a warning will be issued.

Where risks have been reduced to as low as possible (ALAP), they shall be further assessed to determine if they can be reduced to as far as possible (AFAP).

- **Risk Control and Verification:** Risk control measures shall conform to safety principles and take into account generally acknowledged state-of-the-art.

Action plans to eliminate the risk or control/reduce the risk shall be assessed and implemented such that the residual risk associated with each hazard and overall residual risk is acceptable.

The risk associated with the risk reduction/elimination action shall be verified to determine if it has created an alternative or additional risk.

Implement risk reduction/elimination plans shall be verified to ensure that it is effective. The verifications shall be recorded in the Risk Management File.

Each risk shall be weighed to decide if the risk is acceptable and if the risk is residual and if necessary, warnings are given (provided that the residual risk outweighs the medical benefits of the device).

Appropriate solutions shall be selected such that it eliminates or reduces risk as far as possible through safe design and manufacture; where appropriate, takes adequate protection measures and provides information for safety and training where appropriate.

ANNEXURE C

7.3. RISK PLAN TEMPLATE 2

Risk Plan for Business A

- **Purpose and Scope:**
    - **Purpose:** The purpose of this document is to ensure that the risk management activities carried out by Business A are planned and aligned with the relevant standards and regulations.
    - **Scope:** This risk plan is applicable for risk assessments conducted by Business A on the following medical devices:

      (list medical devices)

This risk plan is applicable throughout all life cycle phases of the product inclusive of: planning/design and development, manufacture, packaging and labelling, sales and marketing, distribution, post market and disposal.

- **Device Information**
- **Device Name**

This risk plan is applicable to the following medical devices being manufactured by Business A:

(list medical devices by trade name)

- **Manufacturer**

(List some history about the manufacturer)

- **Device Description**

(Provide a brief description of the medical devices which this risk plan is applicable to. This description should match what is listed in the technical file)

- **Definitions**
    - **Harm** – Injury or damage to the health of people, or damage to property or the environment
    - **Hazard** – Potential source of harm
    - **Life Cycle** – Series of all phases in the life of a medical device from the initial conception to final decommissioning and disposal
    - **Residual Risk** – Risk remaining after risk control measures have been implemented
    - **Risk** – Combination of the probability of occurrence of harm and the severity of that harm
- Abbreviations

None

- Roles and Responsibilities

The risk assessment team of business a shall consist of the following persons:

- The Managing Director;
- General Manager;
- Quality Manager;
- Regulatory Manager
- Group Engineer;
- Other personnel from various departments (sales, production, etc.) as required.
- External consultants (where necessary)

Business A shall ensure that the members of the Risk Assessment Team shall have sufficient knowledge and experience in risk assessment and management techniques and application of the device. Suitable records of competence shall be maintained.

- Applicable Documents:

(Business A to list all applicable standards and regulations used when writing the risk plan here)

- Overview of Procedure for Risk Management:

The risk management process shall be managed in accordance with the following figure as below:

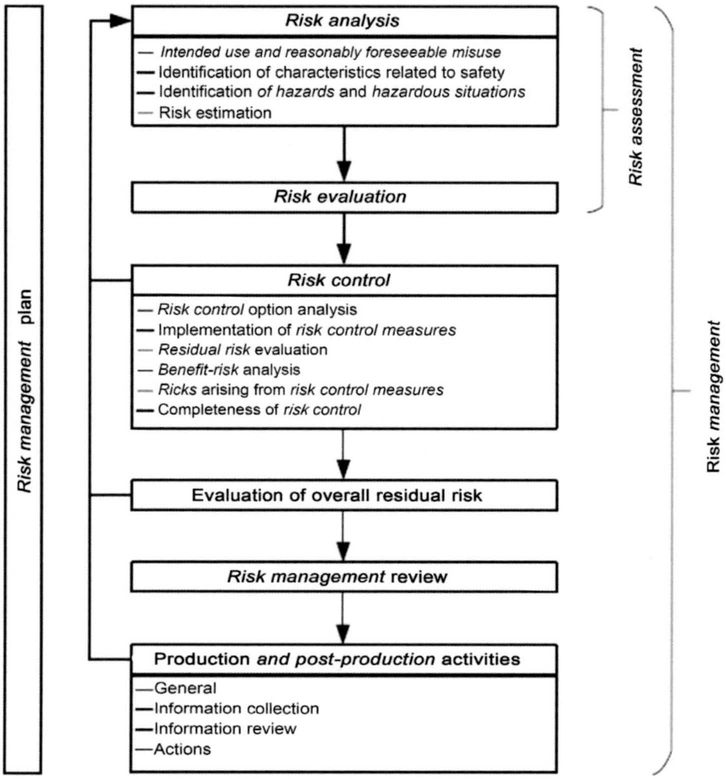

A risk review shall be performed if previously unrecognized hazards or hazardous situations are identified, or otherwise as a minimum, an annual risk review shall be conducted for the product.

In the event of a new or undated risk assessment, a review and update of the General Safety and Performance Requirements Checklist should be conducted.

The Risk Assessment Matrix, Risk Management Report and Risk Management File shall be updated where indications of new or increased risks are obtained from the following:

- Internal non-conformances;
- Customer Complaints;
- Input from Product Development activities during new product development;
- Input from the post-market surveillance system.

- Where major changes are done to the product characteristics;
- Where the class of the medical device is changed due to a new application;
- Where raw materials or packaging materials are changed;
- Where the sterilization process is changed and
- Where raw material suppliers or subcontractors are changed.

Criteria for Risk Acceptability

Estimation/quantification of risks for each hazard shall be based on the worst-case scenario/single fault condition.

Severity rating table:

Severity	Rating	Description
Life-Threatening	10	▪ Death has or could occur
Severe	8	▪ Serious injury, permanent impairment, irreversible injury
Moderate	6	▪ Significant impairment that is either temporary or reversible
Limited	4	▪ Transient / self-limiting illness or injury
None	2	▪ No injury or adverse health consequence anticipated

Risk acceptability criteria:

Probability of Occurrence	Rating	Qualitative Estimate Description	Quantitative Estimate Probability
Continuously Occurring	10	Occurrence is almost inevitable.	≥10% As frequent or more frequent than 1 in 10
Frequent	8	Occurrence is expected to occur repeatedly.	≥1%, but less than 10% As frequent or more frequent than 1 in 100 but Less frequent than 1 in 10
Occasional	6	Occurrence is expected to occur occasionally.	≥ 0.1%, but less than 1% As frequent or more frequent than 1 in 1,000 but Less frequent than 1 in 100
Rare	4	Occurrence is expected to occur relatively infrequently.	≥ 0.01%, but less than 0.1% As frequent or more frequent than 1 in 10,000 but Less frequent than 1 in 1,000
Remote	2	Occurrence is not likely to occur.	less than 0.01% (<1:10,000) Less frequent than 1 in 10,000

Probability of Occurrence of Harm	Severity				
	2 **None** (No injury or adverse health consequence anticipated)	**4** **Limited** (Transient / self-limiting illness or injury)	**6** **Moderate** (Significant impairment that is either temporary or reversible)	**8** **Severe** (Serious injury, permanent impairment, irreversible injury)	**10** **Life-Threatening** (Death has or could occur)
10 **Continuously Occurring** (almost inevitable)	As Low as Possible	As Low as Possible	Unacceptable	Unacceptable	Unacceptable
8 **Frequent** (expected to occur repeatedly)	Acceptable	As Low as Possible	As Low as Possible	Unacceptable	Unacceptable
6 **Occasional** (expected to occur occasionally)	Acceptable	Acceptable	As Low as Possible	As Low as Possible	Unacceptable
4 **Rare** (expected to occur relatively infrequently)	Acceptable	Acceptable	Acceptable	As Low as Possible	As Low as Possible
2 **Remote** (not likely to occur)	Acceptable	Acceptable	Acceptable	Acceptable	As Low as Possible

All risks shall receive attention to determine methods to minimize or eliminate the risk. Where the residual risk cannot be reduced or eliminated and the benefits outweigh the risk a warning will be issued.

Where risks have been reduced to As Low As Possible, they shall be further assessed to determine if they can be reduced to As Far As Possible.

Risk Control and Verification

Risk control measures shall conform to safety principles and take into account generally acknowledged state-of-the-art.

Action plans to eliminate the risk or control/reduce the risk shall be assessed and implemented such that the residual risk associated with each hazard and overall residual risk is acceptable.

The risk associated with the risk reduction/elimination action shall be verified to determine if it has created an alternative or additional risk.

Implement risk reduction/elimination plans shall be verified to ensure that it is effective. The verifications shall be recorded in the Risk Management File.

Each risk shall be weighed to decide if the risk is acceptable and if the risk is residual and if necessary, warnings are given (provided that the residual risk outweighs the medical benefits of the device).

Appropriate solutions shall be selected such that it eliminates or reduces risk as far as possible through safe design and manufacture; where appropriate, takes adequate protection measures and provides information for safety and training where appropriate.

ANNEXURE D

7.4. RISK PROCEDURE TEMPLATE

Purpose and Scope:

- • **Purpose:** The purpose of this procedure is to define the process followed by Business A when performing a product risk assessment on a medical device as per ISO 14971.
- • **Scope:** This procedure is applicable to all risk assessments being conducted on medical devices for Business A.

Definitions and Abbreviations:

- **Definitions:**
    - o **Harm** – Injury or damage to the health of people, or damage to property or the environment
    - o **Hazard** – Potential source of harm
    - o **Life Cycle** – Series of all phases in the life of a medical device from the initial conception to final decommissioning and disposal
    - o **Residual Risk** – Risk remaining after risk control measures have been implemented
    - o **Risk** – Combination of the probability of occurrence of harm and the severity of that harm

The Risk Assessment Team shall consist of the following individuals:

- The Managing Director;
- General Manager;
- Quality Manager;
- Regulatory Manager
- Group Engineer;
- Clinical Evaluators/Healthcare Professionals, etc., as required
- Other personnel from various departments such as sales, engineering, production as required.

The members of the Risk Assessment Team shall have sufficient knowledge and experience in risk assessment and risk management techniques as well as the use and application of the device.

Records of qualification and experience shall be maintained.

Abbreviations:

- ALAP – As low as possible
- AFAP – As far as possible

Applicable Documents/Standards/Regulations

The following documents were used in establishing this procedure within Beier Drawtex Healthcare:

(list any relevant standards and regulations that were consulted when writing the procedure for Business A)

- MDR 2017/745: EU Medical Device Regulation
- ISO 9000 – Quality Management Systems: Fundamentals and Vocabulary
- ISO 9004 – Quality Management Systems: Guidelines for performance improvements
- ISO 13485 – Medical devices – Quality management systems – requirements for regulatory purposes
- ISO 14971 – Medical devices – application of risk management to medical devices
- ISO 24971 – Medical devices – guidance on the application of ISO 14971
- SOR 98-282 – Canadian Medical Device Regulation
- 21 CFR 820 USA Medical Device Regulation
- other territory specific regulations

Procedure for Risk Management:

Risk shall be managed in accordance with the figure below as taken from ISO 14971.

Risk Management System:

Business A shall establish, implement, document and maintain and risk management system as per ISO 14971.

Risk Management File:

The Quality Manager shall establish, implement, document and maintain a Risk Management File. This shall be initiated at the onset of the design and development phase of a device, or for existing devices retrospectively.

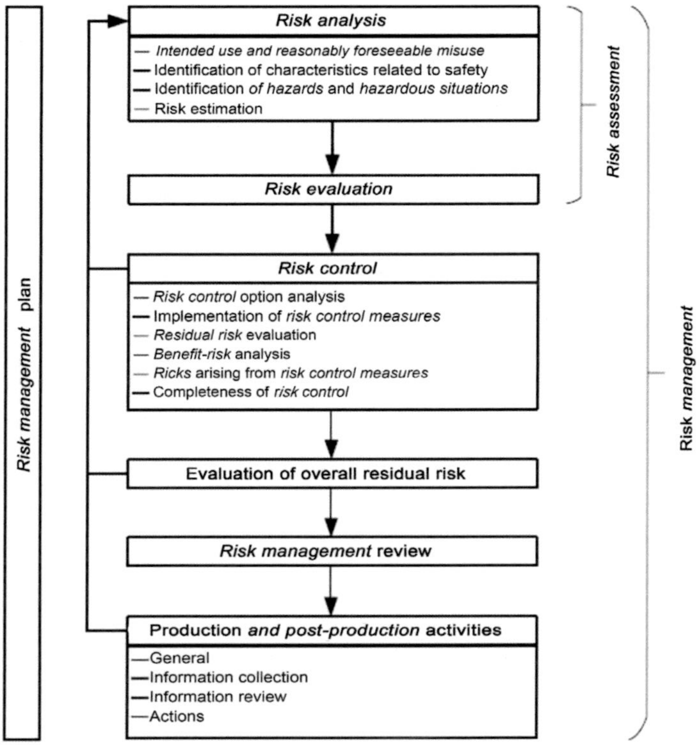

The risk management file shall be a continuous process maintained throughout the entire lifecycle of the device.

The Risk Management File shall at least consist of the following:

- Risk Policy
- Risk Management Plan
- Risk Assessment matrix
- Risk Management Report
- Risk-Benefit Report

Risk Assessment:

Intended use/purpose of the device and related characteristics:

The intended use/purpose of the device and/or accessories and its characteristics shall be defined.

This definition shall be based on the following:

- Required skill and training of the user;
- Ergonomic Factors

- The role the device will play in:
  - diagnosis
  - prevention
  - monitoring
  - treatment or alleviation of disease
  - compensation for injury
  - replacement
  - modification of anatomy, etc.
- Is the device sustaining or life supporting?

Identification and analysis of known and foreseen hazards and harms associated with the device

The following will be considered in the identification of known and foreseen hazards and harms:

- Is the device intended to have contact with the patient or other persons?
- What materials and/or components are incorporated in the device or are used with, or are in contact with the device?
- Is energy delivered to and/or extracted from the patient?
- Are substances delivered to and/or extracted from the patient?
- Are biological materials processed by the device for subsequent re-use?
- Is the device supplied sterile or intended to be sterilised by the user or are other microbiological controls applicable?
- Is the device intended to be routinely cleaned and disinfected by the user?
- Is the device intended to modify the patient environment?
- Are measurements taken?
- Is the device interpretative?
- Is the device intended to control or to interact with other devices or drugs?
- Are there any unwanted outputs of energy or substances?
- Is the device susceptible to environmental influences?
- Does the device influence the environment?
- Are there essential consumables or accessories associated with the device?

- Is maintenance and/or calibration necessary?
- Does the device have a restricted shelf life?
- Are there any possible delayed and/or long-term use effects?
- To what mechanical forces will the device be subjected?
- What determines the lifetime of the device?
- Is the device intended for single use?
- Is safe decommissioning or disposal of the device necessary?
- Does installation of the device require special training?
- Will new manufacturing processes need to be established or introduced?
- Is successful application of the device critically dependent on human factors such as the user interface?
    - o Connecting parts and accessories
    - o Control interface
    - o Display information
    - o Controlled by a menu
- Is the device intended to be mobile or portable?

Estimation/Evaluation/Quantification of risks for each hazard/harm

The Risk Assessment Team shall conduct a risk assessment in accordance with this procedure as well as ISO 14971 and ISO 24971 as a guide.

Estimation and evaluation of risks associated with, and occurring during the intended use and during foreseeable misuse shall be considered.

The worst-case scenario/single fault condition shall be predicted and an evaluation of the risk made.

The Risk Assessment shall be completed taking into consideration the General Safety and Performance Requirements Checklist.

The following risk evaluation criteria shall be used:

Severity rating table

Severity	Rating	Description
Life-Threatening	10	▪ Death has or could occur
Severe	8	▪ Serious injury, permanent impairment, irreversible injury
Moderate	6	▪ Significant impairment that is either temporary or reversible
Limited	4	▪ Transient / self-limiting illness or injury
None	2	▪ No injury or adverse health consequence anticipated

Risk Acceptability Criteria:

Probability of Occurrence	Rating	Qualitative Estimate Description	Quantitative Estimate Probability
Continuously Occurring	10	Occurrence is almost inevitable.	≥10% As frequent or more frequent than 1 in 10
Frequent	8	Occurrence is expected to occur repeatedly.	≥1%, but less than 10% As frequent or more frequent than 1 in 100 but Less frequent than 1 in 10
Occasional	6	Occurrence is expected to occur occasionally.	≥ 0.1%, but less than 1% As frequent or more frequent than 1 in 1,000 but Less frequent than 1 in 100
Rare	4	Occurrence is expected to occur relatively infrequently.	≥ 0.01%, but less than 0.1% As frequent or more frequent than 1 in 10,000 but Less frequent than 1 in 1,000
Remote	2	Occurrence is not likely to occur.	less than 0.01% (<1:10,000) Less frequent than 1 in 10,000

	Severity				
Probability of Occurrence of Harm	**2** **None** (No injury or adverse health consequence anticipated)	**4** **Limited** (Transient / self-limiting illness or injury)	**6** **Moderate** (Significant impairment that is either temporary or reversible)	**8** **Severe** (Serious injury, permanent impairment, irreversible injury)	**10** **Life-Threatening** (Death has or could occur)
10 **Continuously Occurring** (almost inevitable)	As Low as Possible	As Low as Possible	Unacceptable	Unacceptable	Unacceptable
8 **Frequent** (expected to occur repeatedly)	Acceptable	As Low as Possible	As Low as Possible	Unacceptable	Unacceptable
6 **Occasional** (expected to occur occasionally)	Acceptable	Acceptable	As Low as Possible	As Low as Possible	Unacceptable
4 **Rare** (expected to occur relatively infrequently)	Acceptable	Acceptable	Acceptable	As Low as Possible	As Low as Possible
2 **Remote** (not likely to occur)	Acceptable	Acceptable	Acceptable	Acceptable	As Low as Possible

All risks shall be reviewed and assessed to minimize or eliminate the risk. Where risks are reduced to As Low As Possible, they will be further assessed to determine if they can be reduced to As Far As Possible.

Where the residual risk cannot be reduced or eliminated and the overall benefit of the device outweighs the risk, a warning will be issued.

Risk Control:

Risk control measures shall conform to safety principles and take into account generally acknowledged state-of-the-art.

Action plans to eliminate the risk or control/reduce the risk shall be assessed and implemented such that the residual risk associated with each hazard and overall residual risk is acceptable.

The risk associated with the risk reduction/elimination action shall be verified to determine if it has created an alternative or additional risk.

Implemented risk reduction/elimination plans shall be verified to ensure that it is effective. The verifications shall be recorded in the Risk Management File.

Each risk shall be weighed to decide if the risk is acceptable and if the risk is residual. If necessary, a warning is given.

Appropriate solutions shall be selected such that it eliminates or reduces risk as far as possible through safe design and manufacture; where appropriate, takes adequate protection measures and provides information for safety and training where appropriate.

Acceptability of the Benefit Risk Ratio:

Acceptability of the benefit-risk ratio will be determined based on:

- A risk assessment of the above-mentioned list of clinical specifications
- Risk mitigations that consider state of the art
- A calculation of a benefit-risk score

Overall Risk Evaluation:

The Risk Assessment Team shall evaluate the identified risks and risk reduction/ elimination plans. Where applicable, they shall issue a statement that the overall risks of the device are acceptable.

Risk Management Report:

A Risk Assessment Report shall be documented and shall provide traceability for each hazard or harm to the risk analysis, the risk evaluation, the risk implementation and verification of the risk control measures, and that the residual risk/s is acceptable.

Post Production and Post Marketing Information:

The Risk Assessment Matrix, Risk Management Report and Risk Management File shall be updated where indications of new or increased risks are obtained.

This may be obtained from the following:

- Internal non-conformances;

- Customer Complaints;
- Input from Product Development activities during new product development;
- Input from the post-market surveillance system.
- Where major changes are done to the product characteristics;
- Where the class of the medical device is changed due to a new application;
- Where raw materials or packaging materials are changed;
- Where the sterilization process is changed and
- Where raw material suppliers or subcontractors are changed.

ANNEXURE E

7.5. RISK MANAGEMENT REPORT TEMPLATE

Risk Management Report for Business A

- List Product Name
- List Report Author
- List who the report is reviewed and approved by
- List Version and Amendment History

Device Description:

Provide an overview of the description of the device which the risk management report applies too. This could include information such as:

- class of the device
- classification rules
- registrations/certifications
- mode of action
- patents/trade marks
- sterilization information
- packaging/how the device is supplied
- single/multiple use
- contraindications/warnings

Manufacturing Overview:

Provide an overview of the manufacturing process. This could include the following information:

- How the device is manufactured
- In what facility the device is manufactured in
- What type of packaging materials are used
- How many layers of packaging
- Shelf-life claims

Overview of the Risk Assessment:

This section should provide an overview of the businesses risk management process. It should:

- reference to relevant standards
- who the risk assessment team consists of
- experience/expertise of the risk assessment team
- reference to the different stages of the device's life cycle (i.e., design, purchasing, manufacturing, final release, transport and storage, clinical use)
- overview of the risk analysis process, how hazards will be identified, what risk control measures are in place, how the residual risk will be determined, effectiveness of risk control measures and determination of residual risks
- reference to the risk benefit analysis for each hazard
- criteria being used to estimate severity and probability of occurrence of the identified harm for each hazard
- an assigned risk acceptability rating

Risk Summary:

A summary of the risk assessment process should be provided. This should include reference to those risks which have been reduced to ALAP and further to AFAP.

Reference should be made to risks which are deemed to be unacceptable and which have been mitigated, however which require some form of warning, etc.

Reference should be made to those residual risks which are transcribed into labelling/IFUs.

Reference should be made to the risk benefit analysis.

Reference should be made to other supporting documents within the Business, such as:

- post market surveillance reports
- adverse event reports

- sales data
- audit findings
- clinical evidence
- technical documentation
- labelling, etc.

ANNEXURE F

7.6. RISK-BENEFIT REPORT TEMPLATE

Risk Benefit Report for Business A:

- List Product Name
- List Report Author
- List who the report is reviewed and approved by
- List Version and Amendment History

Introduction to the Risk Benefit Report:

This section should cover a brief introduction of the risk benefit report, which standards have been considered and how hazards have been identified.

Device Description:

Provide an overview of the description of the device which the risk management report applies too. This could include information such as:

- class of the device
- classification rules
- registrations/certifications
- mode of action
- patents/trade marks
- sterilization information
- packaging/how the device is supplied
- single/multiple use
- contraindications/warnings

Intended Purpose of the Device:

Discuss the intended purpose of the device in relation to clinical evidence and performance, safety and efficacy of the device.

- Evaluation of the benefit to patients/end users
- Discuss the benefit to the patient and end users

- Evaluation of the clinical risks of the device
- Discuss the clinical risks of the device and any mitigations
- Evaluation of acceptability of the benefit/risk profile
- Provide an overall conclusion indicating that the benefit of the product outweighs any residual risks.

CHAPTER 8

OTHER STANDARDS THAT SHOULD BE CONSULTED WHEN CARRYING OUT RISK MANAGEMENT

Contents

8.1. STANDARDS REFERENCED IN ISO 14971 AND ISO 24971

ISO/IEC Guide 51:2014	Safety aspects — Guidelines for their inclusion in standards.
ISO/IEC Guide 63:2019	Guide to the development and inclusion of aspects of safety in international standards for medical devices.
ISO 9000:2015	Quality management systems — Fundamentals and vocabulary.
ISO 10993-1	Biological evaluation of medical devices — Part 1: Evaluation and testing within a risk management process.
ISO 13485:2016	Medical devices — Quality management systems — Requirements for regulatory purposes.
ISO 14155	Clinical investigation of medical devices for human subjects — Good clinical practice.
ISO 18113-1:2009	In vitro diagnostic medical devices — Information supplied by the manufacturer (labelling) — Part 1: Terms, definitions and general requirements.
ISO 20916	In vitro diagnostic medical devices — Clinical performance studies using specimens from human subjects — Good study practice.
ISO/TR 24971	Medical devices — Guidance on the application of ISO 14971.
ISO 31000	Risk management — Guidelines.
IEC/TR 60513	Fundamental aspects of safety standards for medical electrical equipment.
IEC 60601-1	Medical electrical equipment — Part 1: General requirements for basic safety and essential performance.
IEC 62366-1:2015	Medical devices — Part 1: Application of usability engineering to medical devices.
AAMI TIR 57:2016	Principles for medical device security — Risk management.
GHTF/SG3/N18 2010	Quality management system — Medical devices — Guidance on corrective action and preventive action and related QMS processes.
GHTF/SG5/N4 2010	Post-market clinical follow-up studies.
IEC Guide 120:2018	Security aspects — Guidelines for their inclusion in publications.

IEC 60601-1	Medical electrical equipment — Part 1: General requirements for basic safety and essential performance.
IEC 60601-1-2	Medical electrical equipment — Parts 1–2: General requirements for basic safety and essential performance — Collateral Standard: Electromagnetic disturbances — Requirements and tests.
IEC 60601-1-8	Medical electrical equipment — Parts 1–8: General requirements for basic safety and essential performance — Collateral standard: General requirements, tests and guidance for alarm systems in medical electrical equipment and medical electrical systems.
IEC 60601-2-16	Medical electrical equipment — Parts 2–16: Particular requirements for basic safety and essential performance of hemodialysis, hemodiafiltration and hemofiltration equipment.
IEC TR 60601-4-1	Medical electrical equipment — Parts 4–1: Guidance and interpretation — Medical electrical equipment and medical electrical systems employing a degree of autonomy.
IEC 60812	Failure modes and effects analysis (FMEA and FMECA).
IEC 61010-2-101:2015	Safety requirements for electrical equipment for measurement, control and laboratory use — Parts 2–101: Particular requirements for in vitro diagnostic (IVD) medical equipment.
IEC 61025	Fault tree analysis (FTA).
IEC 61326-2-6	Electrical equipment for measurement, control and laboratory use — EMC requirements — Parts 2–6: Particular requirements — In vitro diagnostic (IVD) medical equipment.
IEC 61882	Hazard and operability studies (HAZOP studies) — Application guide.
IEC 62304:2006 and AMD1: 2015	Medical device software — Software life cycle processes—Amendment.
IEC 62366-1:2015	Medical devices — Part 1: Application of usability engineering to medical devices.
IEC TR 62366-2	Medical devices — Part 2: Guidance on the application of usability engineering to medical devices.
IEC 62502	Analysis techniques for dependability — Event tree analysis (ETA).

IEC 80001-1:2010	Application of risk management for IT-networks incorporating medical devices — Part 1: Roles, responsibilities and activities.
ISO/IEC Guide 63:2019	Guide to the development and inclusion of aspects of safety in International Standards for medical devices.
ISO/DIS 10017	Quality management—Guidance on statistical techniques for ISO 9001:2015.
ISO 10993-1:2018	Biological evaluation of medical devices — Part 1: Evaluation and testing within a risk management process.
ISO 11608-1	Needle-based injection systems for medical use — Requirements and test methods — Part 1: Needle-based injection systems.
ISO 13485:2016	Medical devices — Quality management systems — Requirements for regulatory purposes.
ISO 14155	Clinical investigation of medical devices for human subjects — Good clinical practice.
ISO 15189	Medical laboratories — Requirements for quality and competence.
ISO 15197	In vitro diagnostic test systems — Requirements for blood-glucose monitoring systems for self-testing in managing diabetes mellitus.
ISO 16142-1	Medical devices — Recognized essential principles of safety and performance of medical devices — Part 1: General essential principles and additional specific essential principles for all non-IVD medical devices and guidance on the selection of standards.
ISO 16142-2	Medical devices — Recognized essential principles of safety and performance of medical devices — Part 2: General essential principles and additional specific essential principles for all IVD medical devices and guidance on the selection of standards.
ISO 17511	In vitro diagnostic medical devices — Measurement of quantities in biological samples — Metrological traceability of values assigned to calibrators and control materials.
ISO 17593	Clinical laboratory testing and in vitro medical devices — Requirements for in vitro monitoring systems for self-testing of oral anticoagulant therapy.
ISO/TS 17822-1, In vitro	diagnostic test systems — Qualitative nucleic acid-based in vitro examination procedures for detection and identification of microbial pathogens — Part 1: General requirements, terms and definitions.

ISO 18113	In vitro diagnostic medical devices — Information supplied by the manufacturer (labelling).
ISO/TR 20416	Medical devices — post-market surveillance for manufacturers.
ISO 20776	Clinical laboratory testing and in vitro diagnostic test systems — Susceptibility testing of infectious agents and evaluation of performance of antimicrobial susceptibility test devices.
ISO 20916	In vitro diagnostic medical devices — Clinical performance studies using specimens from human subjects — Good study practice.
ISO 22367	Medical laboratories — Application of risk management to medical laboratories.
ISO 22442	Medical devices utilizing animal tissues and their derivatives.
ISO 23640	In vitro diagnostic medical devices — Evaluation of stability of in vitro diagnostic reagents.

BIBLIOGRAPHY

1. AHRQ. (2009, Dec 18). Agency for Health Research and Quality; Brockton Hefflin; Center for Devices and Radiological Health; Food and Drug Administration. Available from: http://www.healthit.ahrq.gov/portal/server.pt?open=18.ppt (accessed 4 October 2024).
2. American College of Clinical Engineering. (1995). *Guidelines for medical equipment donation*. Plymouth Meeting, PA.
3. American Medical Association. (2003). World Medical Association Declaration of Helsinki: Ethical principles for medical research involving human subjects. *Clinical Review and Education, 2191–2194*.
4. American Medical Association. (2013). World Medical Association Declaration of Helsinki: Ethical principles for medical research involving human subjects. *JAMA, 310*, 2191–2194.
5. Bloom, G., & Temple-Bird, C. (1990). Medical equipment in Sub-Saharan Africa: A framework for policy-formulation. *IDS Research Report No. 19*, WHO/SHS/NHP/90.6, WHO, Geneva.
6. Braybook, J. H. (Ed.). *Biocompatibility assessment of medical devices and materials*. John Wiley & Sons; 1997.
7. Carlson, R. V., Boyd, K. M., & Webb, D. J. (2004). The revision of the Declaration of Helsinki: Past, present, and future. *British Journal of Clinical Pharmacology, 57*, 695–713.
8. Chang, T. (1997). Medical device donations to developing countries: Things to consider. Poster presentation at the 32nd AAMI Annual Meeting, Washington, D.C., USA, June 1997.

9. CMC. (1994). *Guidelines: Medical equipment donations*. Technical Library, FAKT, in: CMC Publication CONTACT 10/94. CMC, World Council of Churches, Geneva.

10. Dionne, G. (2003). Risk management: History, definition and critique. Cirpee – Centre Interuniversitaire sur le Risque, les Politiques Économiques et l'Emploi. Working Paper 13-02.

11. Falcitelli, G. (1994). Biomedical technology and clinical engineering in cooperation programmes with developing countries: Methodological guidelines. In *Proceedings of the regional workshop on health care technology in the Sub-Saharan region* (p. 13), Somerset West, South Africa. South African Medical Research Council, Tygerberg.

12. Ganeshkumar, P. (2002). Basic principles of risk management for medical device design. White paper. Wipro Technologies.

13. Global Harmonization Task Force. (2009, Nov 25). Generating the GMDN process and development. Available from: http://www.ghtf.org/meetings/conferences/8thconference/.../Dillard.ppt (accessed 4 October 2024).

14. GMDN Agency. (2010, Feb 02). Global medical device nomenclature. Available from: http://www.gmdnagency.com/?id=userinfo (accessed 4 October 2024).

15. GMDN Agency. (2010, Jan 10). Global medical device nomenclature: User guide. Available from: http://www.gmdnagency.com/ (accessed 4 October 2024).

16. GMDN Agency. (2010, Jan 14). Global medical device nomenclature. Available from: http://www.gmdnagency.com/?id=userinfo (accessed 4 October 2024).

17. Hadfield, D., et al. The future of risk management in medical devices – How to leverage a PLM backbone and machine learning for a modern, closed-loop approach to risk. Kalypso.

18. Heimann, P., & Poluta, M. A. (1997). Health technology management in the Sub-Saharan region as a prerequisite for optimizing the donor aid intervention process. (In press) WHO, ARA, Geneva.

19. IMDG. (1992). Donating and selling used medical equipment. *Health Devices, 21*(9), 295–297.

20. Issakov, A. (1994). Service and maintenance in developing countries. In: C. W. D. Gruting (Ed.), *Medical devices: International perspectives on health and safety* (pp. 21–28). Amsterdam: Elsevier.

21. Kachienga, M. O. (1992). An overview: Procurement of health care equipment for public health facilities. *Medicus*, 20–23.

22. Khinvasara, T., et al. (2023). Risk management in the medical device industry. *Journal of Engineering Research and Reports, 25*, 130–140.
23. Madani, M. A. (1995). Medical equipment retirement and donation criteria. Presentation to the 30th AAMI Annual Meeting, Anaheim, CA, USA, May 1995.
24. MHRA (Medicines and Healthcare Products Regulatory Agency). (2010a). Medical device alert: DePuy ASR acetabular cups used in hip resurfacing arthroplasty and total hip replacement. *MDA/2010/044*, 25 May 2010.
25. Ministry of Health, Tanzania. (1995). *Guidelines on donations of drugs and medical equipment to the health sector for Tanzania Mainland*. Dar-es-Salaam.
26. Mosenkis, R., (1994). in Grutting, C. *Medical devices: International perspectives on health and safety*. Amsterdam: Elsevier.
27. Sawyer, C. (1997). *Do it by design: An introduction to human factors in medical devices*. FDA.
28. Segura Roca, G. (2020). Risk management in medical devices: Hazard identification and verification of mitigation controls. KTH Royal Institute of Technology, School of Engineering Sciences in Chemistry, Biotechnology and Health.
29. South African Medical Research Council. (1994). *Executive report of the regional workshop on health care technology in Sub-Saharan region*, Somerset West, South Africa, April 1994. Tygerberg.
30. Speer, J. *The definitive guide to ISO 14971 risk management for medical devices*. Greenlight Guru.
31. UNDP/WHO. (1996). *Emergency relief items: Compendium of basic specifications, Volume 2: Medical supplies and equipment*. Geneva.
32. Vroonhoven, J. V. Risk management for medical devices and the new BS EN ISO 14971. *Medical device white paper series*.
33. Wang, B. (1996). The facts on used equipment export. *Biomedical Technology Management*, 64.
34. World Health Organization. (2001). Bulletin of the World Health Organization: World Medical Association Declaration of Helsinki. Ethical principles for medical research involving human subjects.

INDEX

Q

R

U

V

W

X